Handbook of Oil and Gas Piping

Handbook of Oil and Gas Piping

A Practical and Comprehensive Guide

Murali Sambasivan

*School of Business, Taylors University Lakeside Campus,
Subang Jaya, Selangor, Malaysia*

Sekar Gopal

Muhibbah Engineering (M) Bhd, Klang, Selangor, Malaysia

CRC Press
Taylor & Francis Group
Boca Raton London New York Leiden

CRC Press is an imprint of the
Taylor & Francis Group, an **informa** business

A BALKEMA BOOK

CRC Press/Balkema is an imprint of the Taylor & Francis Group, an informa business

© 2019 Taylor & Francis Group, London, UK

Typeset by Apex CoVantage, LLC

Library of Congress Cataloging-in-Publication Data
Names: Sambasivan, Murali, author. | Gopal, Sekar, author.
Title: Handbook of oil and gas piping : a practical and comprehensive guide /
    Murali Sambasivan, School of Business, Taylors University Lakeside Campus, Subang Jaya,
    Selangor, Malaysia, Sekar Gopal, Muhibbah Engineering (M) Bhd, Klang, Selangor, Malaysia.
Description: Boca Raton : CRC Press/Balkema is an imprint of the Taylor & Francis Group,
    an Informa Business, [2019] | Includes bibliographical references and index.
Identifiers: LCCN 2018033893 (print) | LCCN 2018038840 (ebook) |
    ISBN 9780429459757 (ebook) | ISBN 9781138625617 (hardcover : alk. paper)
Subjects: LCSH: Petroleum pipelines—Design and construction | Underground
    pipelines—Safety measures. | Project management.
Classification: LCC TN879.5 (ebook) | LCC TN879.5 S26 2019 (print) |
    DDC 665.5/44—dc23
LC record available at https://lccn.loc.gov/2018033893

Published by: CRC Press/Balkema
                Schipholweg 107c, 2316 XC Leiden, The Netherlands
                e-mail: Pub.NL@taylorandfrancis.com
                www.crcpress.com – www.taylorandfrancis.com

ISBN: 978-1-138-62561-7 (Hbk)
ISBN: 978-0-429-45975-7 (eBook)

Contents

Preface

Someone had to write this book. This book was necessitated by the fact that there are limited comprehensive handbooks available for oil and gas piping projects. The objective of this oil and gas practical piping handbook is to aid oil and gas piping-related construction projects project management teams in understanding the key requirements of the discipline and to equip them with the necessary knowledge and protocols. A comprehensive coverage of all the practical aspects of piping-related material sourcing, fabrication essentials, welding-related items, NDT activities, erection of pipes, pre-commissioning, commissioning, post-commissioning, project management and the importance of ISO management systems in oil and gas piping projects are discussed in this handbook. Most of the oil and gas piping-related project contractors face multi-faceted challenges in completing the projects on time with the required quality and within the cost. Failures such as material rejects, fabrication reworks, NDT failures, preservation-related issues, identification-related issues and excessive punch defects during pre-commissioning are some of the common issues faced by the contractors. This handbook will assist the contractors in ensuring the right understanding and application of related protocols in the project to avoid issues mentioned above. The key highlight of this handbook is that the technical information and the formats provided are practical examples from real-time oil and gas piping projects, and hence application of this information is expected to bring in credentials to the contractors in the eyes of the client as well as simplify the existing operations to an extent. Another important highlight of this handbook is that it covers holistically from the raw material stage till project completion and handover and beyond. This will help the oil and gas piping contractors to train their project management staff to follow the best practices practiced in the oil and gas industry. Furthermore, this piping handbook provides an important indication on the key relative important project-related factors (hard factors) and organizational-related factors (soft factors) to achieve the desired project performance dimensions such as timely completion, within the cost, acceptable quality, safe execution and financial performance. Understanding and taking necessary steps to focus on these important factors specifically implied for oil and gas projects will be an enabler for improved overall project performance. Lastly, role of ISO management systems such as ISO 9001, ISO 14001 and OHSAS 18001 in construction projects is widely known across the industry; however, oil- and gas-specific ISO quality management systems ISO 29001 and project-specific management system ISO 21500 are not widely known in the industry, which are explained in detail in this handbook for the benefit of the oil and gas construction organizations. We acknowledge the support given by our families and respective organizations to complete this handbook.

Murali Sambasivan, PhD
Sekar Gopal, DBA

About the authors

Murali Sambasivan: has more than 18 years of academic experience and 10 years of industry experience. His research interests are in the areas of management science, operations management, supply chain management and other areas of management.

Professor Sambasivan has published more than 80 papers in many international refereed journals such as *International Journal of Production Economics*, *International Journal of Medical Informatics*, *Journal of Biomedical Informatics*, *Journal of Cleaner Production*, *Technovation*, *Computers and Operations Research*, *International Journal of Physical Distribution and Logistics Management*, *International Journal of Project Management*, *Engineering, Construction, and Architectural Management* and many more. He has supervised/co-supervised more than 50 PhD and MSc students. Professor Sambasivan obtained his PhD in management science from University of Alabama at Tuscaloosa, USA, and his degrees (bachelor's and master's) in engineering from India. Has published three study guides for students in the area of statistics and decision making.

Gopal Sekar: has more than 30 years of work experience predominantly in the textile and construction industry. Has worked for more than 10 years in oil and gas construction projects and has managed more than 10 projects. Recently obtained DBA from *Universiti Utara Malaysia*, Malaysia. His thesis was related to project management in all sectors of the construction industry.

List of appendices

List of abbreviations

AC/DC	Alternating Current/Direct Current
AFC	Approved For Construction
API	American Petroleum Institute
ASME	American Society of Mechanical Engineers
ASNT	American Society of Non-Destructive Testing
AWS	American Welding Society
BS	British Standard
BLS	Baseline Survey
BOMBA	Fire and rescue department (Malaysia)
C	Celsius
CAT	Category
CF	Certificate of Fitness
CM	Construction Manager
CNC	Computerized Numerical Control
CoS	Cold Spring
Cr	Chromium
CS	Carbon Steel
CSP	Compact Strip Production
CSWIP	Certification Scheme for Welding and Inspection Personnel
CV	Curriculum Vitae
DC	Document Controller
DCC	Document Control Center
DLP	Defect Liability Period
DN	Diameter Nominal
DNV	Det Norske Veritas (ISO Certification Body)
DO	Delivery Order
DoE	Department of Environment
DOSH	Department of Occupational Safety & Health
DPI	Dye Penetration Inspection
DPT	Dye Penetration Test
DSS	Duplex Stainless Steel
EN	European Norms
FAT	Factory Acceptance Test
GRE	Glass Reinforced Epoxy
HAZOP	Hazard and Operability Study

HSE	Health, Safety and Environment
HT	Hydro Test
HUC	Hook-Up/Commissioning
HV	Hardness Vickers
ID	Identification
IFC	Issued for construction
ISO	International Standards Organization
ITB	Invitation to Bid
ITPs	Inspection and Test Plans
ITRs	Inspection and Test Reports
KPI	Key Performance Indicators
LTCS	Low Temperature Carbon Steel
MC	Mechanical Completion
MCr	Material Controller
Mo	Molybdenum
MPI	Magnetic Particle Inspection
MPT	Magnetic Particle Test
MT	Material test
MTC	Material Test Certificate
MTOC	Master Table Of Contents
N	Number of samples
N_2	Nitrogen
N/A	Not Applicable
NDE	Non-Destructive Examination
NDT	Non-Destructive Test
NPS	Net Pipe Size
O_2	Oxygen
OD	Outside Diameter
OHSAS	Occupational Health, Safety Assurance System
OSC	Online Support Center
PC	Pre-Commissioning
PCN	Personnel Certification in Non-Destructive Testing
PCR	Piping Contractor
PD	Project Director
PE	Project Engineer
PEFS	Process Engineering Flow Scheme
PERT	Program Evaluation Review Technique
P&ID	Process and Instrumentation Diagram
PI	Tag Tag Configurator
PM	Project Manager
PMBOK	Project Management Body of Knowledge
PMI	Positive Material Identification
PMI	Project Management Institute
PMT	Permit *Mesin Tekanan* (Permit for the machine)
PO	Purchase Order
PPE	Personal Protective Equipment
PQR	Procedure Qualification Record

ProM	Procurement Manager
PS	Pre-Spring
PSI	Process Safety Information
PSV	Pressure Safety Valve
PT	Penetration test
PTS	Petronas Technical Standards
PTW	Permit to work
PWHT	Post-Weld Heat Treatment
QA	Quality Assurance
QAF	Quality Assurance Format
QAP	Quality Assurance Procedures
QC	Quality Control
R&D	Research & Development
RFI	Request for Inspection
RT	Radiography Test
SAT	Site Acceptance Test
SBF	Single block female
SDSS	Stability and ductility of steel structure
SIC	Source Inspection Coordinator
SS	Stainless Steel
TPI	Third-Party Inspection
TS	Technical Specification
TUV	Technischer Überwachungsverein (ISO certification body)
UT	Ultrasonic Test
VTOC	Volume Table of Contents
WCS	Weld Control Software
WO	Work Order
WPS	Welding Procedure Specification
WPSR	Welding Procedure Specification Register
WQ	Welder Qualification Test

Overall process flow chart

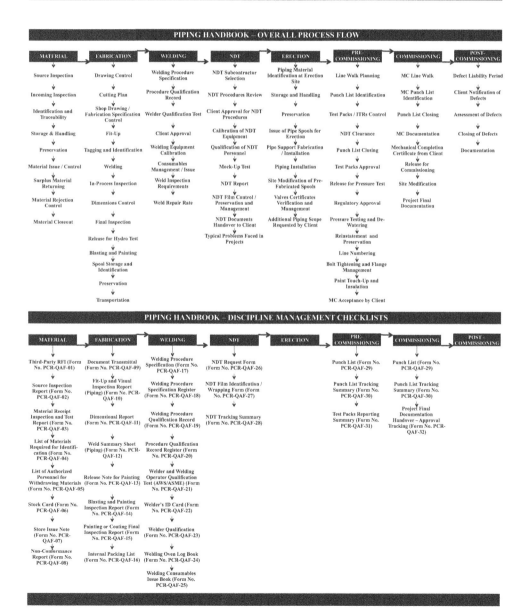

PIPING HANDBOOK – OVERALL PROCESS FLOW

MATERIAL	FABRICATION	WELDING	NDT	ERECTION	PRE-COMMISSIONING	COMMISSIONING	POST-COMMISSIONING
Source Inspection	Drawing Control	Welding Procedure Specification	NDT Subcontractor Selection	Piping Material Identification at Erection Site	Line Walk Planning	MC Line Walk	Defect Liability Period
Incoming Inspection	Cutting Plan	Procedure Qualification Record	NDT Procedures Review	Storage and Handling	Punch List Identification	MC Punch List Identification	Client Notification of Defects
Identification and Traceability	Shop Drawing / Fabrication Specification Control	Welder Qualification Test	Client Approval for NDT Procedures	Preservation	Test Packs / ITRs Control	Punch List Closing	Assessment of Defects
Storage & Handling	Fit-Up	Client Approval	Calibration of NDT Equipment	Issue of Pipe Spools for Erection	NDT Clearance	MC Documentation	Closing of Defects
Preservation	Tagging and Identification	Welding Equipment Calibration	Qualification of NDT Personnel	Pipe Support Fabrication / Installation	Punch List Closing	Mechanical Completion Certificate from Client	Documentation
Material Issue / Control	Welding	Consumables Management / Issue	Mock-Up Test	Piping Installation	Test Packs Approval	Release for Commissioning	
Surplus Material Returning	In-Process Inspection	Weld Inspection Requirements	NDT Report	Site Modification of Pre-Fabricated Spools	Release for Pressure Test	Site Modification	
Material Rejection Control	Dimensions Control	Weld Repair Rate	NDT Film Control / Preservation and Management	Valves Certificates Verification and Management	Regulatory Approval	Project Final Documentation	
Material Closeout	Final Inspection		NDT Documents Handover to Client	Additional Piping Scope Requested by Client	Pressure Testing and De-Watering		
	Release for Hydro Test		Typical Problems Faced in Projects		Reinstatement and Preservation		
	Blasting and Painting				Line Numbering		
	Spool Storage and Identification				Bolt Tightening and Flange Management		
	Preservation				Paint Touch-Up and Insulation		
	Transportation				MC Acceptance by Client		

PIPING HANDBOOK – DISCIPLINE MANAGEMENT CHECKLISTS

MATERIAL	FABRICATION	WELDING	NDT	ERECTION	PRE-COMMISSIONING	COMMISSIONING	POST-COMMISSIONING
Third-Party RFI (Form No. PCR-QAF-01)	Document Transmittal (Form No. PCR-QAF-09)	Welding Procedure Specification (Form No. PCR-QAF-17)	NDT Request Form (Form No. PCR-QAF-26)		Punch List (Form No. PCR-QAF-29)	Punch List (Form No. PCR-QAF-29)	
Source Inspection Report (Form No. PCR-QAF-02)	Fit-Up and Visual Inspection Report (Piping) (Form No. PCR-QAF-10)	Welding Procedure Specification Register (Form No. PCR-QAF-18)	NDT Film Identification / Wrapping Form (Form No. PCR-QAF-27)		Punch List Tracking Summary (Form No. PCR-QAF-30)	Punch List Tracking Summary (Form No. PCR-QAF-30)	
Material Receipt Inspection and Test Report (Form No. PCR-QAF-03)	Dimensional Report (Form No. PCR-QAF-11)	Welding Procedure Qualification Record (Form No. PCR-QAF-19)	NDT Tracking Summary (Form No. PCR-QAF-28)		Test Packs Reporting Summary (Form No. PCR-QAF-31)	Project Final Documentation Handover – Approval Tracking (Form No. PCR-QAF-32)	
List of Materials Required for Identification (Form No. PCR-QAF-04)	Weld Summary Sheet (Piping) (Form No. PCR-QAF-12)	Procedure Qualification Record Register (Form No. PCR-QAF-20)					
List of Authorized Personnel for Withdrawing Materials (Form No. PCR-QAF-05)	Release Note for Painting (Form No. PCR-QAF-13)	Welder and Welding Operator Qualification Test (AWS/ASME) (Form No. PCR-QAF-21)					
Stock Card (Form No. PCR-QAF-06)	Blasting and Painting Inspection Report (Form No. PCR-QAF-14)	Welder's ID Card (Form No. PCR-QAF-22)					
Store Issue Note (Form No. PCR-QAF-07)	Painting or Coating Final Inspection Report (Form No. PCR-QAF-15)	Welder Qualification (Form No. PCR-QAF-23)					
Non-Conformance Report (Form No. PCR-QAF-08)	Internal Packing List (Form No. PCR-QAF-16)	Welding Oven Log Book (Form No. PCR-QAF-24)					
		Welding Consumables Issue Book (Form No. PCR-QAF-25)					

Praise for this work

It is indeed a good initiative to produce this Piping Handbook. This book can serve as a basic reference book that will be of interest to project personnel who are new to piping works to acquaint themselves and as a refresher for experienced project personnel. This book has achieved its objective which sets out to be a general guideline covering all facets of major piping activities including material management, fabrication, installation, pre-commissioning, commissioning and post-commissioning. It addresses important practical aspects and is structured in a way that is easy to understand. In addition, the appendices attached to this book will assist readers in familiarising themselves with common forms used in piping activities. I strongly recommend all project personnel who are involved in piping works to read this Handbook and make the most of it.

Lau Eng Tiong
Yard Operation Manager
Muhibbah Engineering (M) Bhd., Malaysia

The *Handbook of Oil & Gas Piping* is well written. It covers all aspects of piping and piping systems from basic concepts of material management, fabrication, erection, NDT management and commissioning processes. The piping handbook covers the entire field of piping-related construction subjects and follows international piping material specifications, standards, and also the industry standards. It will be a most useful reference book for the oil and gas piping construction project management team members in the construction field as well as in the office.

Shanmugam M. Changany
Project Manager
BASF-Petronas Lemon Grass Project
Muhibbah Engineering (M) Bhd, Malaysia

The *Handbook of Oil & Gas Piping* is well structured and well written. It briefs all practical problems/issues faced in the piping construction field and the solution/s to prevent/avoid such problems/issues. The content of this book ranges from materials to commissioning. This book is well drafted by experienced oil & gas professionals based on their experience in piping projects. This is a great reference book for all personnel involved in piping projects. I recommend to all piping project key personnel to read this book at least once prior to starting their work. The way it is written, it makes the reader to feel easy and gives more confidence in preventing/resolving issues.

Johnson Ambrose
QA/QC Manager
Muhibbah Steel Industries Sdn. Bhd, Malaysia

Material

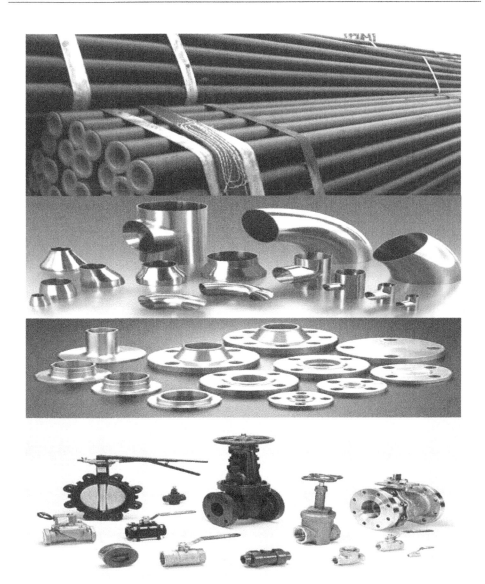

After reading this section, you should be able to understand . . .

1. The importance of oil and gas piping materials and its inspection at source.
2. The process of incoming materials inspection at site.
3. The dos and don'ts of storage, handling and preservation.
4. The process of material issues and control at site.
5. The importance of surplus material management and material rejection control.
6. The procedure for material closeout.

© Can Stock Photo

MATERIAL
↓
Source Inspection
↓
Incoming Inspection
↓
Identification & Traceability
↓
Storage & Handling
↓
Preservation
↓
Material Issue / Control
↓
Surplus Material Returning
↓
Material Rejection Control
↓
Material Closeout

1.0 INTRODUCTION

One of the important items in oil and gas projects is material. It contributes to approximately 50% of the total project cost. Specifically, for oil and gas projects, next to equipment, the important item is piping materials. In this section, oil and gas piping materials and their related management are discussed in detail. Often, material-related issues, such as delay in getting the materials, poor quality of materials, availability of materials, material prices, issues related to material suppliers and poor planning of material procurement, lead to delays, rejects, reworks, additional costs and complaints from clients. Due to the nature of such projects, the piping materials used in oil and gas projects are of high quality for safety and process-related reasons, which need to be properly taken care of through the entire life cycle of the project. Mistakes in materials management in oil and gas projects are very costly and can lead to substantial delays due to limited availability. In this section, all the piping material-related issues are discussed with DOs and DON'Ts to ensure trouble-free project management.

1.1 SOURCE INSPECTION AT VENDOR PREMISES

Source inspection shall be coordinated and scheduled by the source inspection coordinator (SIC), and the inspection shall be performed in two categories:

a) Local purchase orders – the SIC will expedite this inspection through the QA/QC department.
b) Overseas purchase orders – the SIC will expedite this inspection, either with the QA/QC department personnel or by using a third-party inspection agency.

Prior to proceeding with the source inspection, the SIC, as a minimum, shall issue the following key documents to the inspector:

a) Purchase order/work order.
b) Applicable client specification.
c) Client-approved ITPs, procedures, method statements, etc.
d) Client-approved welding procedures, NDT and testing procedures.
e) Client-approved FAT procedures.
f) AFC drawings.
g) Client-approved product data sheets.

The following shall be ensured during the source inspection activity:

DOs	DON'Ts
• Pre-inspection meeting (for long lead delivery items). • Availability of client-approved documents.	• Don't start the work without approved PO/WO. • Don't start the work without client-approved documentation.

(Continued)

DOs	DON'Ts
• Availability of competent resources with vendor. • Availability of materials/status (all MTCs shall be either "ORIGINAL" or "CERTIFIED TRUE COPY" signed and stamped by TPI). • Establish the method of resolving technical queries and application of deviation request. • Establish the method of communication. • Establish the method of reporting progress/schedule/recovery plan/non-conformances. • Establish the inspection and test schedule. • Establish the date of FAT. • Establish the date of final inspection. • Establish the contractual delivery date. • Establish the final documentation requirements. • Establish the logistics methods.	• Don't deploy incompetent inspectors (PCR and TPI) to perform inspection. • Don't deploy inspectors (PCR and TPI) without client approval. • Don't allow the vendor to substitute the materials without client approval. • Don't allow the vendor to proceed from one stage to another stage without client approval (for WITNESS & HOLD POINTS). • Don't allow the vendor to deliver the materials without completion of FAT and closing of punch list. • Don't allow the vendor to deliver the materials without client approval of final documentation, including inspection and release of warranty/guarantee certificates.

Piping contractor (PCR) shall establish the following reference documents in more detail:

Reference Procedures		
PCR-QAP-01	:	Expediting Procedure
PCR-QAP-02	:	Source Inspection Coordination Procedure

For reporting of the activities, the following sample formats shall be used:

Sample Reporting Formats		
Annexure – 1	:	Third-Party RFI (Form No. PCR-QAF-01)
Annexure – 2	:	Source Inspection Report (Form No. PCR-QAF-02)

Note:

In order to prevent issues from clients on source inspection of materials, please invite the client if required for the inspection. This will help to prevent any issues at a later stage.

However, please ensure proper documentation and material availability at vendor premises prior to client approval. A pre-inspection on status prior to client's visit to vendor's premises is advisable.

1.2 INCOMING INSPECTION AT FABRICATION SHOP

Incoming material inspection can be classified and performed as follows:

a) **Permanent materials supplied by client**

   i.   Incoming inspection shall be carried out by material controller (MCr)/storekeeper to verify the dimensions and quantity.

ii. QC inspector shall verify the physical damages and material defects and also review the MTCs for their compliance to relevant standards.

iii. MCr shall notify the status of received, accepted and rejected materials to the client.

Note: Please refer to Control of Customer Supplied Product Procedure for reference (Document No.: PCR-PUP-03)

b) Permanent materials procured by PCR

i. Incoming inspection shall be carried out by material controller (MCr)/storekeeper to verify the dimensions and the quantity.

ii. QC inspector shall verify the physical damages and material defects and also review the MTCs for their compliance to relevant standards.

iii. MCr shall notify the status of received, accepted and rejected materials to the PE/CM/PM/ProM.

c) Consumable materials procured by PCR

i. Incoming inspection shall be carried out by material controller (MCr)/storekeeper to verify the quantity.

ii. QC inspector shall verify the physical damages and material defects and also review the batch certificates for their compliance to relevant standards.

iii. MCr shall notify the status of received, accepted and rejected materials to the PE/CM/PM/ProM.

The following shall be ensured during the incoming material inspection activity:

DOs	DON'Ts
• Check for the availability of documents, like PO, DO, packing list, MTC, batch certificates, data sheets, etc. • For pipes, check for external/internal pitting, dents, bevel damages, pipe end protection, coating damages, internal/external diameter, valve thickness/schedule, seam/seamless, length, pipe number/heat number, material specification, etc. • For fittings, check for external/internal pitting, dents, bevel damages, damages to flange face, application of preservative on the flange face, internal/external diameter, valve thickness/schedule, class/rating, length, pipe number/heat number, material specifications, etc. • For valves, check for type of valve, external/internal damages, bevel damages (for welded valves), damages to flange face, application of preservative on the flange face and internal, length, diameter, class rating/fire rating, valve tag no., name plate details, end protection covers, material specifications, etc.	• Don't perform inspection without required documentation. • Don't release the materials for fabrication/erection without completion of inspection. • Don't forget to invite the client for critical incoming materials inspection at the site.

(Continued)

DOs	DON'Ts
• For gasket, check for type of gasket, external damages, damages to spiral, diameter, thickness, class/rating, damages to insulation sleeves and number of sleeves, washers, material specifications, etc. • For stud bolts, nuts and washers, check for type, external damages, diameter, length, application of preservative on the threads, class, quantity, material specifications, etc. • Use calibrated instruments for inspection. • If deficiency is observed, quarantine the materials and initiate non-conformance reports.	

PCR shall establish the following reference documents in more detail:

Reference Procedures		
PCR-QAP-01	:	Expediting Procedure
PCR-QAP-03	:	Control of Customer Property Procedure
PCR-QAP-04	:	Material Control Procedure

For reporting of the activities, refer to the following sample formats:

Sample Reporting Formats		
Annexure – 3	:	Material Receipt Inspection & Test Report (Form No. PCR-QAF-03)

1.3 IDENTIFICATION AND TRACEABILITY

Identification and traceability can be classified and performed as follows:

a) **Identification and traceability after receiving inspection**

   i. After receiving inspection, all critical materials shall be identified by tagging, marking, color coding, etc., by the store keeper/QC inspector.
   ii. Material color coding.

Material	Color Code
Carbon steel	Green
Stainless steel	Blue

   iii. All pipes shall be properly tagged for traceability purposes. Marker should be chloride, xylene and toluene free.

b) **Identification and traceability of client-supplied material**

   After receiving inspection, all client materials shall be identified by tagging, marking, color coding, etc., by the material controller/QC inspector.

c) **Identification and traceability of surplus material**

After completion of work, all surplus materials shall be identified by marking, color coding, etc., by the material controller.

d) **Identification and traceability during in-process and final inspection**

During the in-process and final inspection activities, all critical materials shall be identified by suitable marking by the QC inspector.

The following shall be ensured for identification and traceability:

DOs	DON'Ts
• Prepare the list of materials that require identification and traceability before receiving the materials. • Prepare the list of critical incoming materials required for the project. • Establish the type of marking, tagging and color coding for identification and traceability.	• Don't issue/release any materials without proper identification and tagging. • Don't release materials which are "ON HOLD" or "REJECTED" for further processing without prior approval from the QC department. • Don't forget to update the material register on issues related to identification and traceability.

PCR shall establish the following reference documents in more detail:

Reference Procedures		
PCR-QAP-05	:	Inspection, Testing, Identification & Traceability Procedure
PCR-QAP-03	:	Control of Customer Property Procedure
PCR-QAP-04	:	Material Control Procedure

For reporting of the activities, refer to the following sample formats:

Sample Reporting Formats		
Annexure – 4	:	List of Materials Required for Identification (PCR-QAF-04)

1.4 STORAGE AND HANDLING

Storage and handling of materials can be classified and performed as follows:

a) **Storage and handling of materials procured by PCR**

All project-related permanent materials and consumables, like welding electrodes, paints, etc., shall be stored separately and in accordance with client/manufacturer's recommendations.

b) **Storage and handling of client-supplied material**

All client-supplied materials shall be segregated, identified and stored properly in a secured area.

The following shall be ensured during storage and handling activity:

DOs	DON'Ts
• All piping materials shall be stored at elevated level (not touching the ground). • Check for the end covers on the pipes. • Check for the protection of flange faces. • Check for the end covers of valves. • Ensure the stainless steel materials are stored separately and protected properly. • Ensure the GRE materials are stored strictly as per vendor recommendation.	• Don't store the pipes and fitting materials directly on the ground. • Don't mix CS, LTCS, SS and DSS materials during storage. • Don't store materials, like valves, flanges, bolts, studs, etc., in open storage areas.

PCR shall establish the following reference documents in more detail:

Reference Procedures		
PCR-QAP-06	:	Handling and Storage of Incoming Materials Procedure

For reporting of the activities, refer to the following sample formats:

Sample Reporting Formats		
Annexure – 5	:	List of Authorized Personnel for Withdrawing Materials (Form No. PCR-QAF-05)
Annexure – 6	:	Stock Card (Form No. PCR-QAF-06)
Annexure – 7	:	Store Issue Note (Form No. PCR-QAF-07)

1.5 PRESERVATION

Preservation can be carried out as follows:

a) **Preservation of equipment**

   i.   Equipment, such as pumps, valves, etc., shall be preserved as per manufacturer's recommendation.
   ii. Rotating parts shall be rotated, lubricated periodically and protected from adverse weather conditions.
   iii. Machined surfaces shall be preserved with removable type rust preventive coatings as recommended by vendor.
   iv. Nitrogen purging shall be maintained as recommended by vendor.

b) **Preservation of piping items**

   i.   Pipes, fittings, valves, etc., shall be preserved to prevent rusting due to weather conditions.
   ii. SS and DSS pipes shall be preserved by covering with canvas.
   iii. GRE materials shall be preserved as per manufacturer's recommendation.

The following shall be ensured during the preservation activity:

DOs	DON'Ts
• Protect all pipes with end caps. • Preserve all valves with proper lubricants periodically. • Preserve all flange faces with suitable lubricants. • Preserve all stud bolts and nuts with suitable lubricants. • Maintain the preservation plan and preservation log sheet.	• Don't expose the materials to heat, moisture, dust and other environmental conditions, which will affect the quality of the materials.

PCR shall establish the following reference documents in more detail:

Reference Procedures		
PCR-QAP-19	:	Preservation Procedure

1.6 MATERIAL ISSUE/CONTROL

Material issue/control shall be performed as follows:

a) All pipes, fittings, valves, gaskets, stud bolts, nuts, etc., shall be controlled by the MCr.
b) MCr shall maintain the summary of the total materials required for the project; the material receipt shall be issued and returned after completion.
c) Materials shall be issued only based on the AFC drawing requirements.
d) Additional/extra materials shall not be issued unless approved by the construction manager/PM.
e) Any materials required for item modification/changes shall be approved by the client, and the details shall be available with the MCr.
f) Materials shall be issued only to the personnel authorized by the PM.
g) Materials shall be issued after the approval of "Store Issue Note."
h) MCr shall update the stock card after every issue.

The following shall be ensured during material/issue control activity:

DOs	DON'Ts
• MCr shall be appointed prior to receiving any materials. • Prepare a secured storage area prior to receiving the material. • Prepare separate storage containers for storing the critical materials before receiving the same. • Issue materials only with "Store Issue Note." • Update stock card on a daily basis.	• Don't issue materials to unauthorized personnel. • Don't issue additional materials without authorization. • Don't interchange the materials received for one drawing/system to another drawing/system. • Don't forget to update the inventory on a daily basis to prevent a stock-out situation.

PCR shall establish the following reference documents in more detail:

Reference Procedures

PCR-QAP-06   :   Handling and Storage of Incoming Materials Procedure

---

For reporting of the activities, refer to the following sample formats:

Sample Reporting Formats

Annexure – 6   :   Stock Card (Form No. PCR-QAF-06)
Annexure – 7   :   Store Issue Note (Form No. PCR-QAF-07)

1.7 SURPLUS MATERIALS RETURNING

Surplus material shall be controlled as follows:

a)   All surplus piping material shall be returned to MCr and stored separately.
b)   MCr shall ensure that all pipes and fittings shall have the heat numbers mentioned in it.
c)   MCr shall record all the returned materials in the stock card.
d)   MCr shall store the usable and scrap materials separately.
e)   MCr shall preserve a copy of all usable materials, so that these materials can be used in other projects.

The following shall be ensured during surplus materials handling activity:

DOs	*DON'Ts*
• Receive surplus materials with proper authorization. • Receive surplus materials with proper identification/heat number. • Report to PM/PD periodically on surplus material details to ensure better control.	• Don't receive damaged materials as surplus materials. • Don't receive any material without heat number. • Don't mix the scrap materials and usable materials.

PCR shall establish the following reference documents in more detail:

Reference Procedures

PCR-QAP-07   :   Surplus Steel and Scrap Materials Handling Procedure

---

For reporting of the activities, refer to the following sample formats:

Sample Reporting Formats

Annexure – 6   :   Stock Card (PCR-QAF-06)

1.8 MATERIAL REJECTION CONTROL

Material rejection control shall be performed as follows:

a) Materials rejected during the incoming inspection shall be quarantined, and non-conformance reporting process shall be initiated.
b) Materials rejected during fabrication/installation shall be removed and returned to MCr. QC inspector shall initiate non-conformance reporting process, if required.
c) MCr shall maintain records of all rejected materials and initiate action to procure replacement materials, if required.

The following shall be ensured during material rejection control activity:

DOs	DON'Ts
• Identify the rejected materials properly. • MCr shall maintain the rejected materials record and follow up until the replacement. • If materials are rejected due to poor workmanship of the subcontractor, the relevant cost shall be back-charged. • Report material rejection reports to PM periodically.	• Don't mix the rejected materials with accepted materials. • Don't rework on the rejected materials without proper authorization.

PCR shall establish the following reference documents in more detail:

Reference Procedures		
PCR-QAP-08	:	Control of Non-Conforming Product, Corrective and Preventive Action Procedure

For reporting of the activities, refer to the following sample formats:

Sample Reporting Formats		
Annexure – 8	:	Non-Conformance Report (Form No. PCR-QAF-08)

1.9 MATERIAL CLOSEOUT

Material closeout activity shall be performed as follows:

a) MCr shall maintain the records for the materials received, issued and returned.
b) MCr shall tally the total materials received and consumed with the remaining stock.
c) MCr shall submit the closeout status to the PM.

PCR shall establish the following reference documents in more detail:

Reference Procedures		
PCR-QAP-07	:	Surplus Steel and Scrap Materials Handling Procedure

For reporting of the activities, refer to the following sample formats:

Sample Reporting Formats		
Annexure – 8	:	Non-Conformance Report (Form No. PCR-QAF-08)

1.10 SUMMARY

Procurement managers of oil and gas piping projects need to ensure that piping-related materials are procured as per contractual specifications from reputed and acceptable vendors for the client. Source inspection with clients will prevent most of the post-procurement issues. Compliance certificates, like material mill certificates, material test certificates and conformance to performance standards, must be thoroughly ensured prior to delivery of materials to the project sites. Most of the oil and gas piping materials are specialized in nature, and hence, comprehensive material procurement planning and a dedicated expeditor is a must. Over-procurement and/or under-procurement must be avoided as it will lead to cost escalation of the projects. Procurement managers must ensure that the materials transportation, delivery and preservation are taken care by the vendors to prevent damages.

Fabrication

After reading this section, you should be able to understand the importance of . . .

1. Drawing controls and shop drawing management at the piping fabrication shop.
2. Cutting plan and fabrication intricacies.
3. Fit-up of pipe spools and its tagging identification.
4. Welding and in-process inspection of pipe spools.
5. Dimensional control and final inspection of fabricated pipe spools.
6. Hydro testing, blasting and painting of pipe spools.
7. Fabricated pipe spools storage, identification, preservation and transportation.

© Can Stock Photo

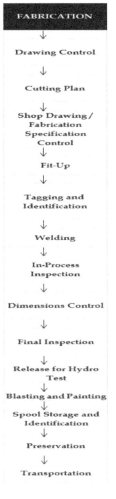

FABRICATION
↓
Drawing Control
↓
Cutting Plan
↓
Shop Drawing /
Fabrication
Specification
Control
↓
Fit-Up
↓
Tagging and
Identification
↓
Welding
↓
In-Process
Inspection
↓
Dimensions Control
↓
Final Inspection
↓
Release for Hydro
Test
↓
Blasting and Painting
↓
Spool Storage and
Identification
↓
Preservation
↓
Transportation

2.0 INTRODUCTION

Fabrication of pipe spools for oil and gas projects highly differs in terms of its technical requirements compared to normal piping fabrication works and/or steel/structural fabrication works. The reason is that pipes are the main transporting element in an oil and gas project for the transfer of oil and gas from one stage to another. Thus, fabrication of such oil and gas pipe spools requires extensive technical capabilities, competencies and experience. History has shown that many of the oil and gas fabricators have failed due to their misunderstanding of the fabrication requirements during the tendering stage, which has led them to perform miserably in terms of timely completion and within the cost in many oil and gas projects. The welding quality requirements, fabrication dimensional quality control requirements, hydro testing requirements, blasting and painting requirements, storage, identification and preservation of pipe spools are all demanding and need special care in the entire fabrication process. This section helps to understand some of these areas in detail to ensure that the fabrication of pipe spools is carried out as per the oil and gas and client requirements.

2.1 DRAWING CONTROL

Drawing control activity shall be carried out as follows:

a) **Control of drawings issued by client**

   i.   All client-supplied drawings shall be maintained separately with a drawing control register maintained by DCC.
   ii. Controlled copy (red stamped) of the drawings shall be distributed with a "Transmittal" to the relevant discipline manager/engineer and subcontractor.
   iii. Any revision/amendment shall be re-issued to the controlled copy holders, and the obsolete revision shall be withdrawn by DCC.

b) **Control of drawings issued by PCR/consultant**

   i.   All PCR/consultant-supplied drawings shall be maintained separately with a drawing control register maintained by DCC.
   ii. Controlled copy (red stamped) of the drawings shall be distributed with a "Transmittal" to the relevant discipline manager/engineer and subcontractor.
   iii. Any revision/amendment shall be re-issued to the controlled copy holders and the obsolete revision shall be withdrawn by DCC.

c) **Control of drawings issued by vendors**

   i.   All vendor-supplied drawings shall be maintained separately with a drawing control register maintained by DCC.
   ii. Controlled copy (red stamped) of the drawings shall be distributed with a "Transmittal" to the relevant discipline manager/engineer and subcontractor.
   iii. Any revision/amendment shall be re-issued to the controlled copy holders and the obsolete revision shall be withdrawn by DCC.

The following shall be ensured during drawing control activity:

DOs	DON'Ts
• Check for the latest revision of drawings from the drawing control register. • Use only red stamped controlled copy (received from DCC) for the fabrication/installation activities. • Return the obsolete drawings to DCC after receiving the latest drawings.	• Don't use the "Information" drawings for fabrication/installation. • Don't receive any unauthorized drawings/sketches that are not issued through DCC. • Don't forget to update the drawings master list daily.

PCR shall establish the following reference documents in more detail:

Reference Procedures		
PCR-QAP-09	:	Document and Data Control Procedure
PCR-QAP-10	:	Design Control Procedure

For reporting of the activities, refer to the following sample formats:

Sample Reporting Formats		
Annexure – 9	:	Document Transmittal (Form No. PCR-QAF-09)

2.2 CUTTING PLAN

The cutting plan shall be prepared for individual AFC piping isometric drawings by the piping engineer. Wherever field-welded joints are shown in the piping isometric drawings, a minimum of 200 mm extra length needs to be considered while making the cutting plan.

Cutting and beveling of pipes may be done by flame cutting, plasma cutting and machining by using CNC machine. The following shall be considered during cutting and beveling:

a) **Carbon steel**

Heat numbers shall be transferred by hard stamping or tagging or by any suitable method prior to blasting and primer application. CS pipe shall be blasted and primed prior to cutting. The cutting and beveling of CS will be done by the flame (oxyacetylene) torch or by CNC machine, and the oxides shall be removed from the surfaces by grinding to bright metal and paint. Grease and rust shall be removed nearer to the bevel ends, if there is any.

b) **Low-alloy steel**

The cutting and beveling of low-alloy steel shall be done by the flame (oxy/acetylene) torch, and after the flame cutting, a minimum of 2 mm of material shall be removed from the cut surface by grinding.

c) **Stainless steel**

The cutting and beveling of stainless steel shall be done by mechanical means using the cutting discs or by plasma cutting (for big size and thick wall) or by CNC machine, and then the surface shall be ground to the bright metal. All the grinding discs to be used for the stainless steel shall be iron free.

d) **Galvanized materials**

Galvanized material shall be cut using cold cutting methods.

The following shall be ensured during cutting and beveling activity:

DOs	DON'Ts
• Pipes for socket weld joints shall be cut to square. • Use SS wire brush for the cleaning of SS materials. • Weld bevels for butt welds of pipes and fittings shall be as per ASME B16.25.	• Don't use CS stands/brackets/supports for cutting SS pipes. • Don't receive any unauthorized drawings/sketches that are not issued through DCC.

PCR shall establish the following reference documents in more detail:

Reference Procedures	
PCR-QAP-11 :	Carbon Steel & Stainless Steel Piping Fabrication and Installation Procedure

2.3 SHOP DRAWING/PIPING SHOP FABRICATION SPECIFICATION

a) The shop fabricator (fabrication shop engineering) shall prepare the necessary detailed shop spool drawings for fabrication from piping drawings on-site under the shop engineer's direction. Shop spool drawings must show the details of field welds, spool number, materials, dimensions, fabrication and the applicable code and procedures.

Shop spool drawings must identify all equipment, lines or spools connecting to that spool. The arrangement or isometric drawing must be shown as a reference.

b) All piping drawings are dimensioned and shown in normal assembly and operating position. When specifically required, thermal expansion requirements and cold springing will be shown on drawings. Cold spring (CoS) and pre-spring (PS) are indicated on the spools with the amount added or deducted. This amount is added or deducted from the dimension, and the pipe shall be fabricated to the dimension.

c) Pipe detail pieces shall be fabricated in accordance with the piece marks shown on the drawings, unless otherwise required for handling and/or shipping. Changes from the piece mark arrangement shown on the drawings shall require approval of the consultant.

d) Preparation of isometric drawings by the shop fabricator is not required.

e)   Piping drawings show the limits of field run piping and the locations of any field welds required between spool pieces. The shop fabricator is not to supply or fabricate that portion of the field runs.

f)   Spool piece numbers shall consist of the isometric line number followed by suffix letters, A, B, C, etc.

g)   Except for ring joint flanges, dimensions are to the centerline of the pipe and the contact face of flanges. This includes the contact face of male and female tongue and groove flanges. Ring joint flanges are dimensioned to the centerline of pipe and to the extreme face of flange and not to the contact surface of the groove.

h)   The shop fabricator (fabrication shop engineering) shall verify all spool drawings and list of materials for accuracy and compliance with the project specifications:

   i.   Quantity of materials correct for the spool
   ii. Description of materials correct as per project specification
   iii. Pipe cut lengths
   iv. All welds numbered for identification on the shop spool drawings and the documents concerned

2.4 FIT-UP

a)   **The fit-up activity can be performed as follows:**

   i.   The weld fit-up shall meet close tolerances as per the WPS.
   ii.   Surfaces to be welded shall be thoroughly cleaned prior to fit-up.
   iii. Temporary tack welding with the suitable size and shape of small pieces (same grade of base material) in the groove may be done.
   iv.   Clamps, holding devices, etc., shall be used to avoid tack welding in the groove. If the type of holding device used for the fit-up requires welding onto the structural steel, such temporary attachments shall be made with the same welding procedure specification as for a permanent weld in the structural member.
   v.   Alignment of equipment shall remain in use until both the root pass and hot pass are completed. Attachment welding onto the pipe, flanges or fittings is not allowed.
   vi.   Tack welding shall be as per the AFC WPS. The extremities of the tack welds shall be dressed by grinding. Wherever bridge tack is specified, the tack welding shall not encroach on the root area. Defective tack welding shall be removed by grinding prior to welding of the joint.
   vii. All tack welds that will form part of the main weld shall be cleaned and ground down to a feather edge at both ends and visually inspected prior to welding the root pass. Tack welds shall be approximately 50 mm long and spaced so that shrinkage forces cannot cause cracking.
   viii. For tack welding, the minimum preheat temperature shall be 50°C and above. Any preheat temperature specified in the WPS for actual welding shall be with a maximum temperature of 150°C.
   ix.   Temporary tack welds shall be removed by grinding or chipping, and the area shall be ground smooth without reduction of wall thickness.
   x.   The minimum distance between the edges of two pressure-containing welds shall be twice the wall thickness of the thicker pressure part. The same requirement

applies to the distance between the non-pressure attachment welds and a pressure-containing weld in order to allow the non-destructive inspection of the pressure-containing weld.

xi. Longitudinal welds in two adjacent pipes should be 180° apart, but in any case shall be separated by at least twice the wall thickness of the thicker pipe.

xii. If the pipe contains a longitudinal weld, this weld shall not be located at the bottom of the pipe after installation. It should be located at least 45° from the bottom of the pipe.

xiii. Where the spool drawings indicate a "Field Weld" in the pipe, the relevant sections shall be supplied 200 mm longer (minimum) with plain ends.

xiv. Misalignment of piping components shall not be more than 1.0 mm or as per the approved WPS. This may be achieved by rotating the pipe. Root misalignment may be corrected by grinding, where the nominal thickness differs. The internal diameter shall be tapered to 1:4.

xv. Where openings for branches are cut in run pipe, all materials that may drop inside the pipe shall be completely removed before the branch line is welded in place.

xvi. Branch connection cutouts shall be properly beveled and accurately matched to form a suitable groove for welding, to enable and permit complete penetration of the welds at all points, resulting in a quality comparable to girth welds, in the same piping system.

xvii. Weld details for branch weld shall be included in the welding procedure specification. Reference is made to the relevant section of ASME B31.3/ASME B31.4/ASME B31.8.

xviii. The centerline of the branch connections shall intersect the centerline of the header, unless otherwise specified in the piping drawing.

xix. Nozzle joints shall be "set-on" type, so that the branch connection abuts the outside surface of the run wall.

b) **After completion of weld joints fit-up, the following inspection shall be carried out prior to commencing welding, and the results shall be within the following specified tolerances:**

i. Flange face alignment

The maximum deviation measured in any direction shall not exceed 2.5 mm/m. When branches are in the same plane and their flanges are also positioned in one plane, the flange facing shall not deviate more than 1 mm from the latter plane in the same direction. Flange faces shall be protected from damages and climatic effect.

ii. Position of bolt holes in flanged piping

The maximum deviation allowed from the required position as measured along the bolt circle is 1.5 mm.

iii. Root gap

For butt welding of all piping components, pipe ends, fittings and welding neck flanges, the root opening shall be as per the approved WPS.

iv. For socket weld, the bottom of the socket and inserted component shall not be in contact (1.5 mm minimum gap is required).

v. Thinning or wall thickness and ovality in bends.
vi. The maximum decrease of wall thickness shall not exceed 10% of the normal wall thickness.
vii. For tolerances on the maximum flattening, refer to ASME B31.3, and where applicable, ASME B31.4/B31.8.

The following shall be ensured during fit-up activity:

DOs	DON'Ts
• Clean all SS weld preparations with acetone. • Use only qualified welders (those qualified in relevant process) for temporary or permanent tag welding. • Check the orientation of longitudinal seam weld during fit-up. • Ensure extra length for field-welded joints.	• Don't make any fit-ups with information drawings. • Don't mix CS/SS welding electrodes/filler wires. • Don't use non-compatible materials as bridging pieces for temporary tag welding.

PCR shall establish the following reference documents in more detail:

Reference Procedures		
PCR-QAP-11	:	Carbon Steel & Stainless Steel Piping Fabrication and Installation Procedure

For reporting of the activities, refer to the following sample formats:

Sample Reporting Formats		
Annexure – 10	:	Fit-Up & Visual Inspection Record (Piping) (Form No. PCR-QAF-10)

2.5 TAGGING AND IDENTIFICATION

Piping spools shall be identified by a detail number, comprising the piping system line number and spool suffix number, which must be weatherproof and painted or stenciled in characters at least 50 mm (2 in.) high.

Piping systems size DN 40 and below (NPS 1½ and below) shall be identified with stainless steel weather- and tear-resistant tags.

The following shall be ensured during tagging and identification activity:

DOs	DON'Ts
• SS binding wire shall be used when fixing tags for SS pipe spools. • For bigger diameter pipes, identification can be marked both inside and outside the pipe.	• Don't release the pipe spool for painting without proper identification tags.

PCR shall establish the following reference documents in more detail:

Reference Procedures		
PCR-QAP-11	:	Carbon Steel & Stainless Steel Piping Fabrication and Installation Procedure

2.6 WELDING

a) The welding consumable shall be controlled as per the manufacturer's recommendation.

b) Welding shall be as per the applicable AFC WPS, and the same shall be performed only by a qualified welder.

c) Work shall not be performed when the weather or degree of protection does not permit satisfactory workmanship. For site welding, adequate shelters shall be provided to protect from wind, rain and moisture. In windy conditions, the pipe ends shall be sealed to prevent through draught.

d) Weld surfaces shall be thoroughly cleaned and dried before welding. Moisture shall be removed to a minimum width of at least 75 mm on either side of the joint by means of blowers, or in exceptional cases, a torch as specified in the AFC WPS may be used.

e) Shielding and purging gas shall be as required in the AFC WPS.

f) Back purging shall be done as given in the AFC WPS. Prior to setup, all hoses and connections shall be checked for leakages. The weld groove containing the purging gas in the chamber between the dams must be sealed from outside.

g) Minimum preheat temperature shall be in accordance with the design code. Where preheating is not required, the actual minimum ambient temperature expected shall be indicated on the WPS.

h) Preheating and inter-pass temperature control requirement shall be as per the AFC WPS. Under no circumstances shall inter-pass temperatures exceed 300°C for the carbon steel or 150°C for the stainless steel. The inter-pass temperature shall be monitored by means of thermos sticks or infrared thermometer.

i) Arc strikes shall be avoided, and if found, shall be ground and the area shall be DPI/MPI examined to ensure defects are removed.

j) Earthing clamps used for the SS welding in direct contact with the SS materials shall have SS contact surface.

k) Proper earthing shall be ensured and condition of cables and attachments shall be periodically examined. Any arcing from the poor connection shall be treated as an arc strike.

l) Welds shall be left as welded and shall not be treated with a flame torch or other mechanical means to change their appearance other than the cleaning and dressing operations specified in the AFC WPS. Welds shall not be peened.

m) After completion of welding, all surfaces shall be cleaned of spatter, burrs and other imperfections.

n) Welding of each weld shall be a continuous operation, with the exception of manually welded root runs for submerged arc welding. In case the welding is to be discontinued, this shall take place after the hot pass is completed.

o) The maximum time that a production weld will remain part welded is 48 hours. The maximum number of heat cycles that will be used in a production weld shall be two. Slow cooling of the weld area shall be ensured. Before continuation of welding, the weld shall be inspected for cracks visually and by MPI. Upon resumption of welding, preheating shall be done in accordance with the approved WPS.

p) Root penetration of welds on orifice flanges shall be ground flush and smooth.

The following shall be ensured during welding activity:

DOs	DON'Ts
• Perform preheating to remove moisture during bad weather conditions. • Nominate welding electrode controller for the control of baking, holding, issue and return of welding electrodes.	• Don't start the welding without the fit-up clearance from the QC inspector. • Don't start the welding without cleaning the fit-up surfaces. • Don't ignore the preheating if it is specified in the WPS. • Don't leave the welding electrodes/filler wires at site after completion of welding activity.

PCR shall establish the following reference documents in more detail:

Reference Procedures		
PCR-QAP-11	:	Carbon Steel & Stainless Steel Piping Fabrication and Installation Procedure
PCR-QAP-12	:	Welding Consumable Control Procedure
PCR-QAP-13	:	Welding Control Procedure

2.7 IN-PROCESS INSPECTION

The following activities shall be performed during in-process inspection:

a) QC inspector shall check the cleaning of fit-ups prior to start of welding.
b) QC inspector shall check the preheating requirement as per WPS.
c) QC inspector shall check the correct WPS, welder and welding consumables required for the joint to be welded.
d) QC inspector shall check the purging requirement for SS welding.
e) Each welder shall check the inter-pass temperature with the tempil stick, and the QC inspector shall monitor the inter-pass temperature.
f) QC inspector shall monitor the requirement of protection due to weather conditions.
g) QC inspector shall record the welder number adjacent to each joint.
h) QC inspector shall monitor the baking, holding, issue and return of welding electrodes.
i) QC inspector shall monitor the use of specific SS tools and wire brushes for the welding of SS materials.
j) QC inspector shall verify the identification tag of each piping spool.

PCR shall establish the following reference documents in more detail:

Reference Procedures		
PCR-QCP-11	:	Carbon Steel & Stainless Steel Piping Fabrication and Installation Procedure
PCR-QCP-12	:	Welding Consumable Control Procedure
PCR-QCP-13	:	Welding Control Procedure

NDE

a) NDE shall be carried out in accordance with the approved procedure of WPS and NDE piping matrix.

b) NDE shall be carried out after 24 hours of welding completion.

c) The NDE technicians shall prepare the NDE reports, which shall be endorsed by the welding inspector.

d) Should NDE examination reveal weld defect, weld repair shall be carried out using the approved WPSR. The repaired weld shall be re-radiographed. Repair shall be limited to a maximum of two attempts. Third repair shall be subject to cutout.

e) The welding inspector shall keep a summary record of the welder's performance. The frequencies of weld repair rate shall be monitored on a weekly basis, based on individual performance and the project overall.

Quality assurance and quality control

a) Inspection records shall be properly recorded and maintained with proper identification for each welded joint.

b) Inspection shall be carried out as per approved procedure, and the results shall be recorded properly.

c) Safety procedures shall be strictly followed in each inspection procedure, and in handling the inspection equipment and work environment.

d) Checklist for spool release:

    i. Conformance to construction and installation specifications.

    ii. Identification and control of materials.

    iii. Personnel are adequately qualified by certification, experience or training.

    iv. Conformance to inspection (QC) and work performance procedures.

    v. Conformance to record-keeping requirements.

    vi. Control of non-conforming items.

2.8 DIMENSIONS CONTROL

The following shall be ensured during dimension control activity:

a) Distance between two weld edges

The minimum distance between the edges of two pressure-containing welds shall be twice the wall thickness of the thicker pressure part.

The same requirement shall apply to the distance between the non-pressure attachment welds and the pressure-containing weld in order to allow non-destructive inspection of the pressure-containing weld.

b) Rotation of longitudinal welds

Longitudinal welds in two adjacent pipes shall be 180° apart, but in any case, shall be separated by at least twice the wall thickness of the thicker pipe.

c) **Location of longitudinal weld**

If the pipe contains a longitudinal weld, this weld shall not be located at the bottom of the pipe after installation. It shall be located at least 45° from the bottom of the pipe.

d) **Dimension control in bending**

If the pipe contains a longitudinal weld, this weld shall be located in the neutral zone of the bend. If the pipe is to be installed horizontally, the longitudinal weld shall be located on the top of the pipe. If one spool is bent in various planes, the longitudinal weld shall be 45° from the top of the pipe.

Shop butt welds shall be located no closer than the smaller of 75 mm (3 inches) or half the nominal pipe diameter to the tangent line. Shop butt welds located the smaller of 150 mm (6 inches) or one nominal pipe diameter to the tangent line shall require 100% radiographic examination unless the inside diameter is accessible for visual inspection.

e) **Dimension control in piping (corrections)**

i.   *For carbon steel that does not require PWHT*

The maximum temperature during the alignment corrections shall be 600°C. Temperature-indicating crayons or contact thermometers shall be used to measure the maximum temperature.

If temperature-indicating crayons are used, a margin of 50°C shall be taken to allow for measurement inaccuracy, i.e., the reading with the crayon shall not exceed 550°C.

Forcing may be applied, if necessary. Cooling in still air shall be applied.

ii. *For carbon steel that requires PWHT*

The maximum temperature during alignment corrections shall be the maximum PWHT temperature.

Temperature-indicating crayons or contact thermometers shall be used to measure the maximum temperature. If temperature-indicating crayons are used, a margin of 50°C shall be taken to allow for measurement inaccuracy, i.e., the reading with the crayon shall not exceed the maximum PWHT temperature minus 50°C.

Only heating with application of local force shall be used, and no quenching shall be used. The area shall either be heated in full compliance with the PWHT procedure or the area that is heated (and corrected) shall be post-weld heat-treated.

iii. *For 0.5 Mo and Cr-Mo steels*

The maximum temperature during the alignment corrections shall be the maximum PWHT temperature. Preferably, contact thermometers shall be used to measure the maximum temperature, but temperature-indicating crayons may also be used.

If temperature-indicating crayons are used, a margin of 50°C shall be taken to allow for measurement inaccuracy, i.e., the reading with the crayon shall not exceed the maximum PWHT temperature minus 50°C.

Forcing may be applied, if necessary. Cooling in still air shall be applied. Random hardness measurements shall be taken and the hardness shall not exceed 248 HV10.

iv.   For austenitic stainless steels

The maximum temperature during the alignment corrections shall be 650°C.

Temperature-indicating crayons or contact thermometers shall be used to measure the maximum temperature. If the temperature-indicating crayons are used, a margin of 50°C shall be taken to allow for measurement inaccuracy, i.e., the reading with the crayon shall not exceed 600°C.

The duration of heating shall be kept as short as possible, and no forcing shall be applied.

f)   Tolerances on length

Length	< 1.5 m	≥ 1.5 m
Distance of any two parallel or crossing centerlines	± 1.5 mm	± 3 mm
Center to flange face		
Flange face to flange face		

g)   Wrinkling tolerance

Wrinkling tolerances shall be as follows:

i.   All wave shapes shall blend into the pipe surface in a gradual manner.

ii.   The maximum vertical height of any wave, measured from the average height of two adjoining crests to the valley, shall not exceed 3% of the nominal pipe size.

iii.   The minimum ratio of the distance between crests as compared to the height between the crests and the valley in between shall be 12 to 1.

h)   Lateral alignment of flanges

i.   For standard flanges, the free insertion of the bolts shall be generally sufficient to demonstrate the acceptable alignment. Lateral alignment may also be checked by laying a straight edge along the outside diameter of the flange. Measurements shall be taken at locations 90° apart around the flange circumference.

ii.   The measured lateral misalignment shall not exceed the following values:

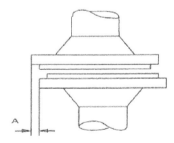

DN	Maximum misalignment
≤ 100	2 mm
> 100	3 mm

   iii. Bolt holes shall straddle the natural centerline unless specified otherwise. The maximum deviation from the required theoretical bolt hole position, as measured along the bolt circle, shall be 1.5 mm.

i) Parallelism of flanges

   i. Flange face alignment shall be checked by measuring the distance between the mating flanges of the pre-assembled joint. Measurements shall be taken around the circumference at equal distances from the centerline (the outside rim of the flange will normally be the most convenient position).

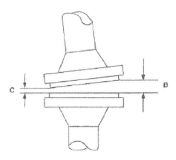

   ii. Parallel misalignment:

- Flange (for ASME B16.50 flanges) B – C = maximum 2.5 mm/m.
- Flange (for ASME B16.47 flanges) B – C = maximum 1.75 mm (regardless of the diameter).

j) Flanged accessories

   i. The individual flange face misalignment from the design plane shall not exceed 2.5 mm/m.

   ii. The misalignment of two flange faces shall not exceed 2.5 mm/m.

   iii. Face alignment of flangeless components (e.g., wafer type control valves, sandwiched between flanges) shall be as follows:

- Flange (for ASME B16.50 flanges) B – C = maximum 2.5 mm/m.
- Flange (for ASME B16.47 flanges) B – C = maximum 1.75 mm (regardless of the diameter).

k) Nozzle faces on static equipment

Alignment of nozzle flange face with the indicated plane shall be within 0.5° in any direction.

l) Flanges connecting to rotating equipment (pumps, compressor, etc.)

The flange face alignment check shall be performed with the bolting inserted loosely, and the acceptance criteria shall be as follows:

Flange Diameter (DN)	Maximum Misalignment at OD of Flange
< 300	0.2 mm
300 to 600	0.3 mm
≥ 600	0.5 mm

PCR shall establish the following reference documents in more detail:

Reference Procedures		
PCR-QAP-14	:	Piping Dimensional Control

For reporting of the activities, refer to the following sample formats:

Sample Reporting Formats		
Annexure – 11	:	Dimensional Report (Form No. PCR-QAF-11)

2.9 FINAL INSPECTION

a) On completion of welding, the QC inspector shall examine the completed welds and the surrounding area to confirm the acceptability of:

    i.   Root penetration and oxidation level (wherever access permits).
    ii. Weld reinforcement.
    iii. Width and size.
    iv. Undercut.
    v.   Overlap.
    vi. Cleanliness (lack of spatter, slag, etc.).
    vii. Stray arcs (arc strikes).
    viii. Satisfactory removal of temporary attachments.

b) Weld finishing shall be as specified in AFC WPS. In the piping/pipeline welding, arc strikes shall not be permitted. If found, the same shall be ground to smooth profile and applicable NDT shall be performed as specified in the relevant specification for acceptance/rejection.

c) Welds shall be examined for smooth and regular profile, and the reinforcement shall blend smoothly with the parent metal.

d) After satisfactory completion of visual inspection, the weld shall be subjected to other NDT methods.

e) After completion of all final inspection activities, the weld summary sheet shall be prepared based on the fit-up, visual and NDT reports.

PCR shall establish the following reference documents in more detail:

Reference Procedures		
PCR-QAP-15	:	Visual Inspection Procedure
PCR-QAP-11	:	Carbon Steel & Stainless Steel Piping Fabrication and Installation Procedure

For reporting of the activities, refer to the following sample formats:

Sample Reporting Formats		
Annexure – 10	:	Fit-Up & Visual Inspection Record (Piping) (Form No. PCR-QAF-10)
Annexure – 12	:	Weld Summary Sheet (Piping) (Form No. PCR-QAF-12)

2.10 RELEASE FOR HYDRO TEST (HT)

After the acceptance of fit-up, visual, NDT and weld summary records, the completed pip-ing spools shall be released for hydro testing (with flanged piping spools). Release for hydro testing shall be given by QC inspector and/or QC engineer.

DOs	DON'Ts
• Check the completion/acceptance of all NDT works. • Check the approval status of all fit-up, visual and NDT reports. • Check the completion of weld summary sheet and NDT percentage. • Check for the completion of all pending repairs.	• Don't release the piping spools if the documents are not signed by client. • Don't release the piping spools if NDT/PWHT has not been completed. • Don't release the piping spools if any repair is pending on that spool.

PCR shall establish the following reference documents in more detail:

Reference Procedures		
PCR-QAP-11	:	Carbon Steel & Stainless Steel Piping Fabrication and Installation Procedure
PCR-QAP-16	:	Air Water Flushing Procedure
PCR-QAP-17	:	Piping Pressure Testing Procedure

For reporting of the activities, refer to the following sample formats:

Sample Reporting Formats		
Annexure – 10	:	Fit-Up & Visual Inspection Record (Piping) (Form No. PCR-QAF-10)

2.11 BLASTING AND PAINTING

After the completion of hydro testing of pre-fabricated piping spools, the piping spools shall be released for blasting and painting. Piping spools that have field-welded joints shall also be released for blasting and painting. Release for blasting and painting shall be initiated by the fabrication supervisor and authorized by the QC inspector-welding and/or QC engineer.

Test panel shall be prepared and witnessed by the client prior to start of painting activi-ties. All blasting and painting activities shall be carried out strictly as per client-approved project-specific procedures.

QC inspector-painting shall inspect the complete blasting and painting operations and perform final inspection prior to release of painted spools.

DOs	DON'Ts
• Prepare paint matrix and obtain approval from client.	• Don't paint any piping spools without identification tag.

- Procure paints after the client approval of paint matrix.
- Perform test panel and obtain approval from paint manufacturer and client.
- Qualify all the blasters and painters before starting the work.
- Calculate the paint wastage properly considering the work location.

- Don't proceed with painting activities without approval of paint matrix and test panel.

PCR shall establish the following reference documents in more detail:

Reference Procedures		
PCR-QAP-18	:	Blasting and Painting Procedure

For reporting of the activities, refer to the following sample formats:

Sample Reporting Formats		
Annexure – 13	:	Release Note for Painting (Form No. PCR-QAF-13)
Annexure – 14	:	Blasting & Painting Inspection Report (Form No. PCR-QAF-14)
Annexure – 15	:	Painting or Coating Final Inspection Report (Form No. PCR-QAF-15)

2.12 SPOOL STORAGE AND IDENTIFICATION

All the painted pipe spools shall be shifted to the pipe spools storage yard and stored as per the erection sequence/system/drawings. All the painted spools shall have proper identification tags at the time of storage. Proper signboards shall be placed to identify the spools for different piping systems. SS pipe spools shall be stored separately and not mixed with CS pipe spools. All painted spools shall not touch the ground during storage. All pipe spool ends shall be covered with pipe caps/polythene, etc., to protect the ingress of foreign materials. The material controller/piping engineer shall maintain a complete list of piping spools that are stored in the storage yard.

PCR shall establish the following reference documents in more detail:

Reference Procedures		
PCR-QAP-19	:	Preservation Procedure

2.13 PRESERVATION

Pipe spool shall be stored at elevated ground level and supported by wooden skid. All pipe spool ends shall be covered with pipe caps/polyethylene, etc., to protect the ingress of foreign materials.

SS/DSS pipe spools shall be stored in the segregated area, where they will not come into contact with foreign materials. They shall be stored off the ground on a nail-free wooden skid. If required, CS/SS pipe spools shall be covered with tarpaulin for long-term storage.

All flange faces shall be protected with wooden blanks. The material controller/QC inspector shall periodically check the storage area.

PCR shall establish the following reference documents in more detail:

Reference Procedures		
PCR-QAP-19	:	Preservation Procedure

2.14 TRANSPORTATION

a)   Transportation of piping spools shall be carried out as per the erection sequence at site.
b)   The piping engineer/supervisor shall prepare the internal packing list and issue it to the material controller for the release of piping spools from the storage yard.
c)   Care shall be taken during loading and unloading so as to not damage the materials as well as the painted surface.
d)   Fabricated pipe spools shall be blocked, strapped or otherwise held in position during shipment and shall be further separated by dunnage as may be necessary to prevent damage.
e)   After receiving the piping spools at site, the site supervisor shall verify the identification tags and endorse the internal packing list that shall be returned to the material controller.
f)   Piping spools shall not be shifted to site without the knowledge of the material controller. After receiving the internal packing list from the site supervisor, the material controller shall update his spool storage summary sheet.

For reporting of the activities, refer to the following sample formats:

Sample Reporting Formats		
Annexure – 16	:	Internal Packing List (Form No. PCR-QAF-16)

2.15 SUMMARY

Construction managers and piping construction engineers of oil and gas piping projects together with the QA/QC discipline engineers should ensure that proper control and care is taken so that all the piping construction activities, such as fit-up, welding, dimensions control, NDT, blasting and painting, are carried out with due diligence as per the approved project specifications and designs. Any errors and mistakes in construction need to be resolved with the help of the QA/QC team prior to inviting clients for acceptance inspection. Involvement of clients during fabrication inspection is a must and will pave the way for further processes. One of the perennial problems faced in oil and gas piping projects during the construction stage is the poor documentation works carried out by the construction team. Project managers and construction managers need to ensure proper attention is paid by all who are involved in the project in this area.

Welding

After reading this section, you should be able to understand the importance of . . .

1. Welding procedure specification (WPS), procedure qualification record (PQR), welder qualification test (WQT).
2. Client approvals for WPS, PQR and WQT.
3. Welding equipment calibration, consumables management and issue controls.
4. Welding inspection requirements and weld repair rates.

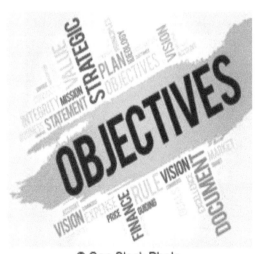

© Can Stock Photo

WELDING

↓

Welding Procedure Specification

↓

Procedure Qualification Record

↓

Welder Qualification Test

↓

Client Approval

↓

Welding Equipment Calibration

↓

Consumables Management / Issue

↓

Weld Inspection Requirements

↓

Weld Repair Rate

3.0 INTRODUCTION

Among the fabrication activities in piping fabrication for oil and gas projects, the foremost fabrication activity is welding. Understanding the piping material to be welded, selection of the right welding machine, usage of appropriate welding procedures, qualified welders, good quality welding consumables and their management are of paramount importance in welding activity. Despite the technical advancements and automation in the welding processes over the last few decades, the need for manual attention and manual control of welding and its related inspection of piping in oil and gas projects still exists and will continue to prevail. Many of the oil and gas PCRs have learned costly lessons due to poor-quality welding and extensive repairs that had to be carried out in the fabricated pipe spools. Since competent welders for oil and gas process piping welding are becoming scarce, it is vital to understand the various technical requirements of oil and gas piping related to welding to ensure fewer repairs, reworks and issues that are likely to delay and add cost to the project. In this section, some of the key aspects of oil and gas piping related to welding are covered in detail.

3.1 WELDING PROCEDURE SPECIFICATION (WPS)

The welding engineer shall prepare the list of welding procedures required for the project based on the piping bill of materials. The welding engineer shall give due consideration to the existing WPSs that are available with the PCR. The welding engineer shall thoroughly study the project scope of work and project specification for welding before compiling the required WPS for CS, SS and DSS materials. The welding engineer shall prepare a weld summary sheet and submit it to the client for review/approval.

PCR shall establish the following reference documents in more detail:

Reference Procedures		
PCR-QAP-20	:	Development of Welding Procedure

For reporting of the activities, refer to the following sample formats:

Sample Reporting Formats		
Annexure – 17	:	Welding Procedure Specification (Form No. PCR-QAF-17)
Annexure – 18	:	Welding Procedure Specification Register (Form No. PCR-QAF-18)

3.2 PROCEDURE QUALIFICATION RECORD (PQR)

After the approval of the WPS summary, the welding engineer shall proceed with the welding procedure qualification. The welding engineer shall arrange for the raw materials, welding consumables, welders, NDT subcontractor and TPI. The welding engineer also shall arrange for the mechanical testing of a test coupon at a client-approved lab.

The welding engineer shall arrange the following calibrated instruments prior to starting the PQR welding:

i.   AC/DC digital clamp meter (min. 2–3 nos.)
ii. Digital infrared thermometer (min. 2–3 nos.)
iii. Digital vernier caliper (1 no.)
iv. High/low gauge (1 no.)
v.   Weld inspection gauge (1 no.)
vi. Stop watch (2–3 nos.)
vii. Steel measuring tape (5 meters) (2 nos.)

The welding engineer shall ensure the following equipment/instruments are calibrated prior to starting the PQR welding:

i.   Welding machines
ii. Electrode baking oven
iii. Electrode holding oven
iv. Flux holding oven
v.   Electrode quivers
vi. Argon gas regulators

DOs	DON'Ts
• Procure full-length pipe for PQR to have proper material identification. • Ensure the availability of MTCs while receiving the materials. • Procure additional quantity of pipes and welding consumables for PQR to meet any PQR failure. • Issue work orders to NDT subcontractor, TPI and laboratory before starting the PQR activities. • If the original PQR is to be submitted to client at the end of the project (with final documentation), prepare an additional original set of PQR (welding, NDT and mechanical testing reports) for the PCR's future project requirement. • Arrange all calibrated instruments (required for welding PQR).	• Don't use inexperienced welders for welding PQR. • Don't start the PQR without having all the abovementioned equipment and instruments. • Don't call third party in advance if all arrangements are not ready for PQR.

PCR shall establish the following reference documents in more detail:

Reference Procedures		
PCR-QAP-20	:	Development of Welding Procedure

For reporting of the activities, refer to the following sample formats:

Sample Reporting Formats		
Annexure – 19	:	Welding Procedure Qualification Record (Form No. PCR-QAF-19)
Annexure – 20	:	Procedure Qualification Record Register (Form No. PCR-QAF-20)

3.3 WELDER QUALIFICATION TEST (WQT)

After the client's approval of the WPS, a WQT shall be performed by the welding engineer/welding inspector. The welding engineer shall check the previous performance record/qualification of the welders prior to the WQT.

Inexperienced welders shall not be given any test. The welding inspector shall strictly inspect the joints visually after weld completion and the welding engineer shall interpret the radiograph strictly to avoid poor-quality welders. The welding engineer shall exercise appropriate control to scrutinize the welders so that during production the weld repair percentage is at a minimum.

The following items shall be considered while issuing a contract to the fabrication subcontractor:

a) Only one test will be given to any welder. Penalty clause will be applicable for repeated WQT.
b) A minimum charge per test shall be mentioned in the welding subcontractor's contract.
c) After passing the WQT, the welder should work until the completion of the project. If a welder leaves the project, a penalty shall be charged to the subcontractor.
d) Individual welder repair percentage shall not exceed the acceptable limit specified for the project/client. If the weld repair percentage exceeds the limit, all the repair costs (equipment, consumables, NDT charges, etc.) shall be charged to the subcontractor.
e) Due to the weld repairs, if any penalty joints are to be radiographed/repaired, the cost related to this NDT/repair also shall be charged to the subcontractor.

DOs	DON'Ts
• Check whether the contract/work order is issued to the fabrication subcontractor, NDT, TPI and mechanical testing lab prior to starting the WQT. • Check the calibrated instruments/equipment area available prior to WQT.	• Don't start the PQR without having sufficient welding consumables, argon gas, etc.

Note: Please involve client representative if required to avoid delays.

PCR shall establish the following reference documents in more detail:

Reference Procedures		
PCR-QAP-21	:	Welder & Welding Operator Qualification Test Procedure

For reporting of the activities, refer to the following sample formats:

Sample Reporting Formats		
Annexure – 21	:	Welder & Welding Operator Qualification Test Certificate (AWS/ASME) (Form No. PCR-QAF-21)
Annexure – 22	:	Welder's ID Card (Form No. PCR-QAF-22)
Annexure – 23	:	Welder Qualification (Form No. PCR-QAF-23)

3.4 CLIENT APPROVAL

All PQRs and WPSs shall be approved by the client prior to starting the WQT.

3.5 EQUIPMENT CALIBRATION

The following equipment/instruments shall be calibrated and made available prior to start of PQR and WQT:

a) Welding machines
b) Electrode baking oven
c) Electrode holding oven
d) Flux holding oven
e) Electrode quivers
f) Argon gas regulators
g) AC/DC digital clamp meter
h) Digital infrared thermometer
i) Digital vernier caliper
j) Steel measuring tape (5 meter)

3.6 CONSUMABLE MANAGEMENT/ISSUE CONTROL

All welding consumables shall be inspected by the QC inspector upon receipt, and after acceptance, shall be stored as per type and size. Electrodes required for baking shall be issued to electrode controller and shall be baked as per manufacturer's recommendation. The details of baking and holding shall be entered into the welding oven logbook. Baked electrodes shall be issued to the welder based on request from the piping engineer/supervisors. Electrodes returned from the site shall be re-baked as per the original baking requirements and the details recorded in the logbook.

Filler wires shall be issued in the original containers and shall be stored properly at the site.

DOs	DON'Ts
• Issue baked electrodes only in quivers.	• Don't issue filler wires in loose condition.
• Perform color coding for the first baking and re-baking.	• Don't leave the welding electrodes and filler wires at site after completion of
• Ensure all quivers are returned to store after completion of each day's work.	each day's work.

PCR shall establish the following reference documents in more detail:

Reference Procedures		
PCR-QAP-12	:	Welding Consumables Storage and Handling Procedure

For reporting of the activities, refer to the following sample formats:

Sample Reporting Formats		
Annexure – 24	:	Welding Oven Log Book (Form No. PCR-QAF-24)
Annexure – 25	:	Welding Consumables Issue Book (Form No. PCR-QAF-25)

3.7 WELD INSPECTION REQUIREMENTS

Weld inspection shall be carried out as per the project specifications and the international standards (AWS D1.1/ASME Section IX). Visual inspection shall be 100%, and other NDE requirements shall be as per the client-approved NDE matrix. Any visual defect notified by the welding inspector shall be repaired immediately, and the re-inspection shall be done by the welding inspector. After completion of weld visual inspection, the welding inspector shall prepare the visual inspection report and submit it to the QC engineer for the compilation of the weld summary sheet.

PCR shall establish the following reference documents in more detail:

Reference Procedures		
PCR-QAP-15	:	Visual Inspection Procedure

For reporting of the activities, refer to the following sample formats:

Sample Reporting Formats		
Annexure – 10	:	Fit-Up & Visual Inspection Report (Piping) (Form No. PCR-QAF-10)

3.8 WELD REPAIR RATE

Welder performance shall be controlled and monitored on a day-to-day basis based on the NDT reports, either by using an Excel sheet or welding software. Based on the individual performance rate, the overall weld repair percentage for the project shall be calculated on a weekly and monthly basis and submitted to PCR top management/client. In general, the overall weld repair rate for the project shall be as follows:

> To include weld repair rate as per client recommendations

a)   For piping – < 0.5% (length wise)
b)   For structure – < 1.0% (length wise)

If any specific project requirement to calculate the weld repair percentage is based on joint or film or by any other method, it shall be resolved with the client prior to starting the production welding.

3.9 SUMMARY

For any oil and gas piping project, the project manager, construction manager and QA/QC manager must understand the importance of this area. Depending on the size and complexity of welding required, the project manager and QA/QC manager must try to deploy a qualified and experienced welding engineer for the project to take care of this area. As the complete oil and gas piping construction is based on welding, the right selection of welders, the welding equipment, the right consumables and the management of consumable issue controls need to be managed at site. Wrong welding procedures, consumables and equipment may result in delay as well as costly expenses. Another important thing is that care should be taken to ensure the weld rejection rate acceptable limits. If the client insists on too stringent weld repair rate acceptable limits, then it is going to be a really difficult task for the contractor to achieve this. Construction managers and piping construction engineers of oil and gas piping projects, together with the QA/QC discipline engineers, should ensure that only industry norms are applicable for the project.

NDT

After reading this section, you should be able to understand the importance of . . .

1. NDT subcontractors selection.
2. NDT procedures and the requirement of client's approval.
3. NDT personnel qualification and calibration of NDT equipment.
4. Mock-up NDT tests and NDT reports management.
5. NDT film control, preservation and management.
6. Process of handing over the NDT documents to clients.

© Can Stock Photo

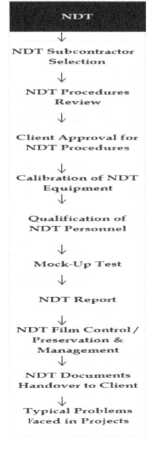

NDT
↓
NDT Subcontractor Selection
↓
NDT Procedures Review
↓
Client Approval for NDT Procedures
↓
Calibration of NDT Equipment
↓
Qualification of NDT Personnel
↓
Mock-Up Test
↓
NDT Report
↓
NDT Film Control / Preservation & Management
↓
NDT Documents Handover to Client
↓
Typical Problems Faced in Projects

4.0 INTRODUCTION

Upon completion of oil and gas piping spools, the fabricated pipe spools need to be tested for their integrity to be used in oil and gas mediums. The most common technique for testing these pipe spools is NDT (non-destructive testing). Some of the most common NDT techniques are radiography test (RT), ultrasonic test (UT), magnetic particle test (MPT) and dye penetration test (DPT). Many other advanced NDT methods are available as well, but the above are the most used tests in oil and gas projects. As such, fabrication facilities are not encouraged to perform these tests internally due to conflict of interest. Most of oil and gas clients insist the fabrication contractors must have independent third-party NDT companies to carry out these tests. Thus, NDT activities have become the most important testing activity in the piping fabrication landscape. Many a time, poor performance of these independent NDT companies has led to project delays, additional testing costs and other associated issues in the project. Thus, it is of real importance to understand the various key aspects of NDT scope of activities and its requirements to ensure that these activities are carried out with due diligence. In this section, the most common NDT techniques and the critical points to be looked at for successful NDT work completion are explained in detail.

4.1 NON-DESTRUCTIVE TESTING (NDT) SUBCONTRACTOR SELECTION

- NDT subcontractor selection should be done well in advance, before the fabrication works start at the project site. It is not advisable to start working on the NDT subcontractor selection work after the fabrication works have started. This will create accumulation of work and delay in carrying out the required NDT work, including identification and traceability issues.
- A minimum of two subcontractors need to be selected for the project based on the scope of NDT work involved in the project. This is to ensure issues, if any, related to the single NDT subcontractor, such as NDT personnel shortage, payment-related issues, etc.
- The potential NDT subcontractor should have a PETRONAS-approved license to carry out NDT work in the project. Care should be taken to check the scope of work mentioned in the PETRONAS license. Absence of PETRONAS license and NDT-related scope of work in the subcontractor's current license will lead to rejection by PETRONAS/client representatives. In some cases, PETRONAS license renewals have not been done by the subcontractors, which is again an issue. In a nutshell, the selected subcontractor should possess a valid PETRONAS license with the required NDT scope of work.
- It must be ensured that the subcontractor has the required NDT equipment and manpower. If not, it will lead to uncertainty in terms of continuous engagement and on emergency needs.
- A check and balance on NDT quoted prices versus market prices versus budget available for the project needs to be worked out. This is to avoid cost overruns due to the NDT scope of work. If the budget allocated is less than the anticipated scope of work, the PM must be informed about the possible cost overruns in this area to avoid cost-related issues later.
- NDT subcontractors who can cover all the NDT scope of work, such as RT/UT/MT/PT/PMI/hardness testing and PWHT, should be selected. If the NDT requirements are very

minimal in nature, then, on a case-by-case basis, this can be relaxed. Otherwise and for bigger projects, it is always advisable to have subcontractors who can cover the entire scope of work related to NDT. This will help to avoid delays and time/effort spent in getting the approval from clients to engage NDT subcontractors.

- NDT subcontractors should be finalized at least one month before the start of actual fabrication work. This is to ensure on-time submission of all the related documents to the client and to facilitate on-time starting of NDT work at the site. Some subcontractors may not provide their NDT procedures if the work order is not issued to them. If fabrication works are started before selection of the subcontractors, it will delay the work.
- The selected subcontractor must submit their company profile, company's track record, previous experience in similar kind of projects and similar scope of work, etc., to avoid surprises later.
- The duration of employment of the NDT subcontractor must be specified in the work order clearly, and the same must be monitored and controlled to avoid cost overruns.

4.2 NDT PROCEDURES

- Based on the NDT scope of work, the NDT procedures must be identified.
- Identified procedures must be obtained from NDT subcontractors and reviewed internally for their correctness prior to submission to client.
- The identified NDT procedures must be approved by the client prior to start of fabrication work.
- The minimum procedures required are RT/UT/MT/PT/PMI/hardness testing and PWHT.
- The selected subcontractor for NDT work must submit the identified procedures within a week of obtaining the work order from the PCR.

4.3 CLIENT APPROVAL FOR NDT PROCEDURES

- Prior to starting NDT work, all the identified NDT procedures must be approved by the client. Comments, if any, by the client during the approval process must be resolved prior to the start of NDT work.
- The client must approve the submitted NDT procedures within two weeks to avoid delay in approvals.
- Close monitoring of NDT approvals by the client is essential for the timely starting of NDT operations in the project.

4.4 CALIBRATION OF NDT EQUIPMENT

- The NDT subcontractor must submit the calibration register and calibration certificates to the PCR together with the NDT procedures prior to mobilization of NDT personnel and equipment to the site.
- The NDT equipment must be provided with calibration status identification for client's verification.
- The client's approval must be obtained for NDT equipment calibration records to avoid issues at a later stage.

- Calibration frequency and expiry of the NDT equipment must be monitored to ensure timely renewal and fitness for use.
- Traceability of the NDT equipment calibration certificates to national and international standards must be ascertained.

4.5 QUALIFICATION OF NDT PERSONNEL

- The NDT subcontractor must submit the CV, qualifications and experience of NDT personnel along with procedures upon receiving the work order from the PCR.
- The PCR must review the CV and qualification certificates of the NDT team and submit them to the client for review and approval.
- Any issues related to the qualifications required for the NDT personnel as against the contractual requirements with the client must be resolved prior to carrying out NDT activities at the site.
- The following minimum qualification requirements for NDT work need to be ascertained.

RT	CSWIP/PCN Level 2
UT	CSWIP/PCN Level 2
MT	CSWIP/PCN Level 2
PT	CSWIP/PCN Level 2
PMI	Adequate practical experience
Hardness test	Adequate practical experience
PWHT	Adequate practical experience

- ASNT-qualified NDT technicians shall be engaged in the projects only if they are approved by the client.
- The PCR must ensure that the NDT personnel who are engaged have valid trade certificates to conduct the specified NDT work at the site.
- Authenticity of the CSWIP/PCN/ASNT certificates possessed by the NDT personnel must be checked, ascertained and communicated to the client, if necessary, to avoid complications later.

4.6 MOCK-UP TEST

- Prior to start of NDT activities in the project, approved procedures and client-specific requirements need to be checked and ascertained.
- Mock-up test must be witnessed by the client as well as NDT subcontractor's Level 3 inspectors.
- Mock-up test related reports must be signed by the PCR's representative, client representative and subcontractor's level three inspector.
- Records of mock-up test must be maintained for future reference.

4.7 NDT REPORT

- Depending on the size, complexity of the project and scope of NDT work, a NDT coordinator needs to be appointed for the project to take care of the NDT-related activities.

If the size of the project is small and the extent of NDT activities is also small, then the QC engineer (welding) can carry out this task.

- The primary roles of a NDT coordinator are as follows:

    1. Raise NDT requests to the NDT subcontractors for the completed welding jobs.
    2. Maintain NDT request tracking and completed NDT reports.
    3. Monitor NDT subcontractors' attendance and time sheet.
    4. Maintain NDT tracking register for 5%, 10% and 100% NDT scope of work based on client and contractual requirements.
    5. Receive the NDT films from the NDT subcontractors and submit them along with the reports to the client.
    6. Ensure approval of client for NDT reports and films.
    7. Compile and submit the NDT reports to QC engineer for test pack compilation.

- The NDT coordinator must keep a copy of all the NDT reports for traceability and reconciliation.
- The tracking register maintained by the NDT coordinator should have details related to NDT failures, penalty, re-shoot, acceptance and rejection.
- Once the NDT-related activities are completed, a copy of the NDT summary shall be issued to the contracts department to settle the NDT subcontractor's claim and related disputes, if any, in the project. For reporting of the activities, the following sample formats shall be used:

Sample Reporting Formats

Annexure – 26   :   NDT Request Form (Form No. PCR-QAF-26)

4.8 NDT FILM CONTROL/PRESERVATION AND MANAGEMENT

- The PCR must ensure that the selected subcontractor is provided a dark room facility to inspect and store the NDT films.
- The NDT coordinator and NDT subcontractor must ensure that the completed NDT films are stored in air-conditioned storage rooms in proper boxes.
- All the films must have proper identification details, such as report number, drawing number, joint number, welder number, weld size, pipe schedule and the results on the wrapping paper of the film.
- The NDT coordinator shall be the custodian for the NDT film control and its management.
- Client review and approval is required before final handover of the NDT films to the client.
- All NDT films shall be packed in aluminium boxes for final handover to the client. Box sizes must be discussed with and finalized by the client.
- Only NDT reports shall be kept by the PCR for reference. There is no need to keep duplicates and/or copies of NDT films unless deemed necessary. For reporting of the activities, the following sample formats shall be used:

Sample Reporting Formats

Annexure – 27   :   NDT Film Identification Wrapping Slip (Form No. PCR-QAF-27)

4.9 HANDING OVER NDT DOCUMENTS TO CLIENT

•   All NDT documents, such as:

1.   NDT procedures
2.   NDT reports
3.   NDT technicians' qualification
4.   NDT films,
     shall be submitted to the client as part of the final documentation for review/
     approval and clearance.

•   The PCR shall maintain a copy of the final documents related to NDT management,
     either in soft or hard form or both for future reference.

4.10 PROBLEMS FACED IN PROJECTS

The following problems faced in previously completed oil and gas projects need to be understood and avoided in future projects:

•   The PCR did not use any monitoring software to capture the data related to weld sum-
     mary, welders' performance, NDT tracking or weld repair percentage, which led to
     delays, complaints from clients, etc. To avoid similar problems in future projects, it is
     advisable to use "weld control software (WCS)" for proper and independent monitoring
     and control.
•   If the client has weld control software, it is for the client's use, not for the PCR. Hence, it
     is always better for the PCR to have the software and not depend on the client to furnish
     the data.
•   A dedicated QC engineer needs to be allocated to control the weld-related performance
     in each and every project.
•   For RT completion, RT balance, RT rejection, RT acceptance, RT repair and RT for penalty
     joints, the PCR must have the required statistics, rather than relying on the client's data.
•   Many a time, clients did not update their weld/NDT data in their software, which led to
     wrong reporting and understanding about the progress in the project.
•   The PCR maintained the NDT/weld tracking data mainly in Excel sheets. This is not a good
     practice. It leads to a lot of mistakes and reconciliation later, which is time-consuming and
     costly. For reporting of the activities, the following sample formats shall be used:

Sample Reporting Formats

Annexure – 28	:	NDT/Weld Tracking Form (Form No. PCR-QAF-28)

4.11 SUMMARY

The QA/QC manager and welding engineer of the oil and gas piping project must ensure
that proper NDT subcontractors are selected for the job, and their technicians must have the
right qualifications and experience. Many of the oil and gas projects have learned valuable
lessons in this area. A comprehensive plan for proper NDT management from the planning

phases of the project is very essential for the smooth and successful execution. NDT films and NDT report management are very important to ensure client satisfaction and to control costs in this area. QA/QC managers must also think of using the right software to ensure that the NDT data are captured progressively depending on the complexity and size of the project. If the size of the project is big, using Excel sheets to manage NDT and weld-related data is not advisable. Project managers must establish a clear budget for NDT work in the project and must carry out monitoring to avoid cost overruns in the project.

Erection

After reading this section, you should be able to understand the importance of . . .

1. Pipe spools and piping materials identification at the erection site.
2. Storage, handling and preservation of piping materials and spools at erection site.
3. Pipe spools issue controls at site.
4. Pipe supports fabrication, installation and pipe spools erection at site.
5. Site modification of pre-fabricated pipe spools and their management at erection site.
6. Valves certification verification and their installation.
7. Additional piping scope of works requested by client and its management.

© Can Stock Photo

ERECTION
↓
Piping Material Identification at Erection Site
↓
Storage and Handling
↓
Preservation
↓
Issue of Pipe Spools for Erection
↓
Pipe Support Fabrication / Installation
↓
Piping Installation
↓
Site Modification of Pre-Fabricated Spools
↓
Valves Certificates Verification and Management
↓
Additional Piping Scope Requested by Client

5.0 INTRODUCTION

Upon completion of all the fabricated spools and ensuring their conformance to NDT testing, the piping spools are sent to the project site for erection. Care should be taken during transportation of the fabricated spools to avoid damage and loss. Apart from this, if the transported spools do not have proper identification tagging, it will create many problems at the project site. Thus, pipe spools should be provided with proper identification tagging and necessary preservation measures at the laydown area of the project site. Similarly, necessary care should be taken while installing the pipe spools along with valves and fittings. Specifically, oil and gas valves are very special and have to meet specific quality criteria. The project management team should ensure that the valves procured have proper test and compliance certificates for the intended processes. Many oil and gas contractors face improper valve selection and end up with costly re-purchase and conflicts with the clients. This needs to be avoided. Further, scope changes to piping and site modification in oil and gas project construction site(s) are very common phenomena. Capturing the variation in the scope of work and the additional scope of work will certainly help the contractors to make some reasonable profits. In this section, some of the above aspects are discussed in detail.

5.1 PIPING MATERIAL IDENTIFICATION (DURING FABRICATION AND INSTALLATION)

a)   All piping materials shall have a proper spool identification mark. For pipes DN100 and larger, marking shall be executed in white block capitals of minimum 19 mm height. For smaller diameter, marking shall be done with permanent markers/paints.

b)   No marking is allowed inside the pipe. Pipes shall be marked longitudinally. Tagging shall be done on fabricated piping items by means of tacking a small stainless steel/zinc plate onto the material or tying with steel wire around the material with a plate tag attached to the wire.

c)   Carbon steel material (general guideline)

   • All pipe spools shall be arranged properly and tagged with a zinc plate.

d)   Stainless steel material (general guideline)

   • All pipe spools shall be arranged properly and tagged with a stainless steel plate.
   • All carbon steel pipe stands shall be covered with plywood/rubber to avoid contamination between these two different materials.

5.2 STORAGE AND HANDLING

a)   **Storage of completed spools**

   i.   A designated area shall be allocated for all completed spools. The area should be feasible for locating the spools.
   ii.   All open ends of piping spools must be covered with plastic caps until integration/offshore final hook-up.
   iii. All hook-up spools shall be tagged as per IFC isometric.
   iv.   Fabricated completed spools shall not be laid down directly on the ground.

v.   All the spools must be segregated according to piping system/service/size/test package.

vi.   Spools shall also be segregated by material.

vii. The supervisor must ensure all spools have proper tagging according to test pack, ISO and line number prior to loading.

viii. Proper care shall be taken of the spools until they are erected on topside.

ix.   Segregated areas for all the surplus and scrap material shall be assigned separately as per requirements.

b) **Material handling**

i.   When pipes are being loaded or unloaded, each length or spool shall be handled individually. Pipes, fittings and flanges shall be handled with care and protected from impact.

ii.   Coated pipe spools and pipe support shall be handled with wide non-abrasive belts, etc.

iii. Raw materials/finished spools shall be handled/lifted/loaded/unloaded/transferred by forklift.

iv.   Forklift blade must be protected against contamination when handling stainless steel material.

v.   Fabricated spools shall be transferred by trailers for painting and erection.

vi.   All lifting gears shall be checked by HSE prior to use.

vii. The relevant PCR's material control procedures must be referred to.

viii. Handling of carbon steel

  •   All tools used for carbon steel material shall not be mixed with exotic material.

ix.   Handling of stainless steel

  •   All tools for stainless steel material shall be separately stored from carbon steel items. Contact with carbon steel components must be avoided.

5.3 PRESERVATION

a) Shop-fabricated pipe must be sectionalized to fit into a box of size 12.0 m × 3.0 m × 3.0 m in accordance with shipping length.

b) When heat-treated pipes of this size cannot be contained within the furnace, the fabricator shall inform the consultant of the maximum size.

c) Before pipe spools are transported from the fabricating shop, the following preparation and protection shall be carried out:

i.   *Flanged ends*

  Coat flange faces with a grease type rust preventive, and cover faces with plywood, hardboard, steel or equivalent covers, which must be securely wired or bolted to the flange proper on at least four bolt holes. All ferrous flange facings shall be coated with anti-rust material.

ii.   *Beveled and plain ends*

  Heavy gauge plastic cover completely sealing the opening and protecting the pipe ends (or consultant-approved alternate) must be provided.

iii. *Threaded ends*

Plastic or metal plugs for female threads and plastic caps for male threads must be provided.

iv. *Machined surfaces*

Machined surfaces must be coated with a grease type rust preventive and provided with a securely fastened wooden or metal cover.

v. *Socket weld ends*

Socket weld ends must have a plastic plug.

Note: Ring type joint flanges shall have grooves packed with grease prior to attaching flange protective covers.

d) The inside of each fabricated pipe spool shall be thoroughly cleaned of all welding spatter or icicles, sand, scale or other foreign debris.
e) Small pieces shall be boxed or wired together to avoid loss in transit.
f) Loading and handling shall be done with reasonable care and details braced where required to prevent damage during transit.
g) Piece marking shall be considered for each spool.
h) Spools must be identified by their line number and spool suffix, which shall be painted or stenciled as characters at least 50 mm high. The size of characters shall be suitable to size of the pipes. Painted numbers shall be located and repeated as necessary, in such a manner that any spool can be easily identified without turning or lifting it.
i) Stamping of piece mark numbers on carbon steel pipes with steel die low stress stamps is permitted, if done lightly and carefully, so as to minimize notch effect and such that the marking will not cause cracks or reduce the wall thickness below the minimum allowed. Stamping of alloy pipes and stainless steel is not permitted without the consultant's approval. Waterproof paint or paint protected with a clear waterproof varnish shall be used on alloy pipes and stainless steel.
j) The welder's identification symbol shall be marked adjacent to each weld. Metal stamping of carbon steel pipes is permissible. On alloy pipes, the marking shall be as follows:

- Heat numbers shall be transferred onto spools and onto spool drawings.
- Weld identification symbols shall be added to spool drawings.
- The paint for marking shall be non-zinc type to prevent zinc contamination.

5.4 ISSUE OF SPOOLS FOR ERECTION

Spool issue/control shall be performed as follows:

a) All spools shall be controlled by the MCr. Spool list shall be prepared and monitored.
b) The MCr shall maintain a summary of the total spools required for the project and the spool receipt, issued and returned, if any.
c) Spools shall be issued upon supervisor's request.
d) Any spools requiring modification/changes shall be approved by the client, and the details shall be available with the MCr.

e)   Spools shall be issued only to the personnel authorized by the PM.
f)   Spools shall be issued after the issue of "Spool Issue Note."
g)   The MCr shall update the spool list after every issue.

The following shall be ensured during spool/issue control activity:

DOs	DON'Ts
• MCr shall be appointed prior to receiving any spools. • Prepare a secured storage area prior to receiving the spools. • Issue spools only with "Spool Issue Note." • Update spool list on daily basis.	• Don't issue spool to unauthorized personnel. • Don't interchange the spools received for one drawing/system to another drawing/system.

PCR shall establish the following reference documents in more detail:

Reference Procedures		
PCR-QAP-22	:	Store Management Procedure

For reporting of the activities, refer to the following sample formats:

Sample Reporting Formats		
Annexure – 6	:	Stock Card (Form No. PCR-QAF-06)
Annexure – 7	:	Store Issue Note (Form No. PCR-QAF-07)

5.5 PIPE SUPPORT FABRICATION/INSTALLATION

a) General

i.   Stainless steel piping shall be adequately protected from contamination.

ii. Piping spools, valves, piping components, equipment and pipe support shall be installed as per approved drawings.

iii. The PCR inspector shall check and verify that spools and pipe supports are installed at proper elevation and correct tie-in locations.

iv. Erection of spools shall be carried out area by area as per piping layout and iso-metric drawings.

v.   Soft slings shall be used for lifting or handling of pipe spools or supports after completion of blasting and painting to minimize coating damage.

vi. Cleaning/flushing of spools shall be done prior to erection.

vii. Pipe spools shall be installed without forcing to avoid under-stress of the line. All flange-to-flange surface alignment must be recorded both before and after welding.

viii.   Pre-fabricated spools shall be shifted to site carefully; care shall be taken while handling and stacking spools to prevent any possible damage.

ix.   All pipe openings shall be sealed before, during and after erection to prevent the ingress of moisture and foreign matter. Threaded ends shall be plugged and sealed by waterproof grease tape or custom-made plastic caps or plugs. End cap or cover shall be placed on the open valve ends.

x.   NDT and other inspection shall be completed for the field joints and recorded properly.

xi.   Final corrections and modifications shall be made to fabricated spools in order to allow for stress-free installation, which should include adjustment to pipe support, adjustment of flanges where free spaces are available in bolt holes, cutting and re-welding or introduction of additional field joints or fit-up pieces.

b)   Pipe support

i.   Piping shall be suitably supported during erection to prevent any sagging and mechanical stress on pipe spool.

ii.   Pipes shall be arranged on support in such a manner that the lengths of the over-hang on either side are approximately equal.

iii.   Fabrication and installation of pipe supports will commence prior to installation of pipe spools. If temporary supports are used, they must be able to take the load off the pipe.

iv.   Pipes should be wedged properly at two points on either side to prevent the pipes from rolling. Wedges are by means of wood or steel materials.

v.   All stainless steel pipes shall be rested on carbon steel supports that have rubber or ply sheet covering the points of contact.

vi.   All temporary pipe supports shall be removed after all permanent supports are in place. Area shall be ground flush and the surface shall be touched up and painted.

vii.   Flanges shall be checked to assure that no strain is induced on the equipment. Pipes shall be removed and corrected if not in correct alignment.

viii.   Alignment corrections shall not be made while pipes are connected to the equipment.

5.6 PIPE INSTALLATION

a)   Cleaning of piping before erection

Before erection, all pre-fabricated spool pieces, pipes, fittings, etc., shall be cleaned inside and outside by suitable means. The cleaning process shall include removal of all foreign matter, such as scale, sand, weld spatter chips, etc., by wire brushes, cleaning tools, etc., and blowing with compressed air/or flushing out with water. Special cleaning requirements for some services, if any, shall be as specified in the piping material specification, isometric or line list.

b)   Piping routing

No deviations from the piping route indicated in drawings shall be permitted without the consent of the company site representative.

Pipe-to-pipe, pipe-to-structure/equipment distances/clearances as shown in the drawings shall be strictly followed, as these clearances may be required for the free expansion of piping/equipment. No deviations from these clearances shall be permissible without the approval of the company site representative.

In case of fouling of a line with other piping, structure, equipment, etc., the matter shall be brought to the notice of the company site representative and corrective action shall be taken as per the representative's instructions.

c) Slopes

Slopes specified for various lines in the drawings/P&ID shall be maintained by the contractor. Corrective action shall be taken by the contractor in consultation with the company site representative wherever the contractor is not able to maintain the specified slope.

d) Flange connections

While fitting up mating flanges, care shall be exercised to properly align the pipes and to check the flanges for trueness, so that faces of the flanges can be pulled together, without inducing any stress on the pipes and the equipment nozzles. Extra care shall be taken for flange connections to pumps, turbines, compressors, cold boxes, air coolers, etc. The flange connections to the equipment shall be checked for misalignment, excessive gaps, etc., after the final alignment of the equipment is over. The joint shall be made up after obtaining approval of the company site representative.

Temporary protective covers shall be retained on all flange connections of pumps, turbines, compressors and other similar equipment until the piping is finally connected, so as to avoid any foreign material from entering the equipment.

The assembly of a flange joint shall be done in such a way that the gasket between these flange faces is uniformly compressed. To achieve this, the bolts shall be tightened in a proper sequence. All bolts shall extend completely through their nuts but not more than ¼ inch.

e) Vents and drains

High point vents and low point drains shall be provided as per the instructions of the company site representative, even if these are not shown in the drawings. The details of vents and drains shall be as per piping material specifications/job standards.

f) Valves

Valves shall be installed with spindle/actuator orientation/position as shown in the layout drawings. In case of any difficulty in doing this or if the spindle orientation/position is not shown in the drawings, the company site representative shall be consulted and work must be done as per the representative's instructions. Care shall be exercised to ensure that unidirectional valves bearing "Flow direction arrow" on the valve body are installed with the arrow pointing in the correct direction. If the direction of the arrow is not marked on such valves, this shall be done in the presence of the company site representative before installation.

Fabrication of stem extensions, locking arrangements and interlocking arrangements of valves (if called for) shall be carried out as per drawings/instructions of the company site representative.

g) **Instruments**

Installation of in-line instruments, such as restriction orifices, control valves, safety valves, relief valves, rotameters, orifice flange assembly, venturi meters, flowmeters, etc., shall be a part of the piping erection work.

Fabrication and erection of piping up to first block valve/nozzle/flange for installation of offline instruments for measurement of level, pressure, temperature, flow, etc., shall also be a part of piping construction work. The limits of piping and instrumentation work will be shown in drawings/standards/specifications. Orientations/locations of takeoff for temperature, pressure, flow, level connections, etc., shown in drawings shall be maintained.

Flushing and testing of piping systems that include instruments mentioned above and the precautions to be taken are covered in flushing, testing and inspection of piping. Care shall be exercised and adequate precautions need to be taken to avoid damage and entry of foreign matter into instruments during transportation, installation, testing, etc.

h) **Line-mounted equipment/items**

Installation of line-mounted items, like filters, strainers, steam traps, air traps, desuperheaters, ejectors, sample coolers, mixers, flame arrestors, sight glasses, etc., including their supporting arrangements, shall form part of the piping erection work.

i) **Bolts and nuts**

The contractor shall apply moly coat grease mixed with graphite powder (unless otherwise specified in piping classes) to all bolts and nuts during storage, after erection and wherever flange connections are broken and made up for any purpose whatsoever. The grease and graphite powder shall be supplied by the contractor within the rates for piping work.

5.7 SITE MODIFICATION OF PRE-FABRICATED SPOOL

The final correction and modifications made to fabricated spool in order to allow for stress-free installation should include adjustment to pipe support, adjustment of flanges where free spaces are available in bolt holes, cutting and re-welding or introduction of additional field joints or fit-up pieces.

The following requirements shall apply:

a) All modifications to the formerly approved design specifications must be done in accordance with the code of construction, which must be in compliance with standards and codes.

b) If a system is to be modified or altered, the new design specifications must be verified by engineering calculations approved by the engineering consultant and client.

c) Modifications must be performed by qualified personnel.

d) All modifications must be recorded. The contractor must complete and submit the following for any modification performed:

    i.   Technical queries

    ii. Redline markup

    iii. Examination and inspection records

5.8 VALVES CERTIFICATES

BS EN 10204:2004 is British Standard that is related to different types of inspection documents for steel metal products. There are four types for EN 10204: 2.1, 2.2, 3.1 and 3.2. The latter two types are mostly used in steel pipe and plate products. Here, the major differences between the two types are given.

a)   What is EN 10204?

When the steel line pipe and fittings are purchased, the manufacturer should release the mill test certificate (MTC) to the buyer. It contains all the specifications of the steel pipe products, including dimensions, sizes, weight, chemical composition, mechanical strength, heat treatment status, test result, traceability, etc. This information ensures the quality of ordered steel products, and tells the buyer what situations could be applied for engineering purposes. So the certificate standard for the MTC is generated, EN 10204 is European Standard for the inspection of documents for steel products, including steel line pipe, fittings, steel plate, valves, sucker rods, etc. For certifying the results of the specific test, ensure that it complies with client's order.

b)   Differences between EN 10204–3.1 and 3.2 for the MTC

EN 10204 contains four types of documents: 2.1, 2.2, 3.1 and 3.2. Types 2.1 and 2.2 are validated by the manufacturer. Types 3.1 and 3.2 are not only validated by the manufacturer.

Type 2.1	–	The content is statement of compliance with the order. Validated by manufacturer.
Type 2.2	–	Statement of compliance with the order, with indication of results of non-specific inspection. Validated by manufacturer.
Type 3.1	–	Statement of compliance with the order, with indication of results of specific inspection. Authorized inspection representative by the manufacturer, but is independent of the manufacturing department.
Type 3.2	–	Statement of compliance with the order, with indication of results of specific inspection. Authorized inspection representative by the manufacturer, independent of manufacturing department and either the buyer's authorized inspection representative or the inspector designated by regulations.

c)   EN 10204–3.1 type quality certificate

It requires the manufacturer to show the actual test result for the steel pipes on sale. According the related standard sampling methods, 3.1 MTC requires the test agency shall be an independent party; mill has no rights to revise the test results.

If a steel pipe manufacturer passes the audit for the ISO 9001 from a certification agency of the European Union, then this manufacturer has the qualification to release the EN 10204–3.1 MTC. The buyer's information shall be specified on the MTC quality certificate; one buyer needs one MTC quality certificate. If the manufacturer does not pass ISO 9001 or the ISO 9001 certificate is not from the European Union Inspection Agency, then the manufacturer does not have the right to release MTC of EN 10204–3.1.

In this case, the manufacturer shall apply for the EN 10204–3.2 quality certificate from a third-party inspection agency.

d) **EN 10204–3.2 type quality certificate**

EN 10204–3.2 certificate is the most restricted standard level for steel pipe products. It indicates the certification shall be additionally countersigned and verified by an independent third party related to all the tests. The 3.2 certificate costs are more than the costs of 3.1. It is essential that the certification is done by the third party in addition to the manufacturer's in-house testing department. Third-party inspector, or the personal buyer, or the government's representative also has the right to verify the test results.

EN 10204–3.2 certificate must be released by the inspection agency authorized by the European Union. The material must be certified by a third-party and confirm that the material ordered is as specified in the PO. The quality certificate EN 10204–3.2 shall specify the name of the manufacturer and the buyer.

e) **EN 10204 certificate related steel pipe MTC in following specs:**

- Chemical composition
- Mechanical strength
- Impact strength
- Tensile strength
- Hardiness test
- Bend test
- NDE
- Visual and dimensions
- Hydro test
- NDT test
- Magnetic test
- Pipe end
- Corrosion HIC status
- Heat number

5.9 ADDITIONAL PIPING SCOPE REQUESTED BY CLIENT

The instruction to the contractor may state that the scope of changes are not expected to have any impact on project timing or cost. In the event of a disagreement, the contractor can respond with a variation request or notice of a time extension or claim. Depending on the contract, the contractor may file a claim for time-related costs.

Managing changes and variations in construction projects:

a) Within the collaborative culture of the project, any participating organization that identifies a change in the contracted scope of works is expected to report that change and provide evidence to support it.
b) The change order should be submitted to the appropriate consultant, contractor or subcontractor for project impact assessment and approval before it goes to the owner.
c) Project documents that have already been registered and issued to the team should not be updated and circulated until the change order has been approved by the owner.

Erroneous document revisions and distributions could compound project risk, leading to schedule delays and cost overruns.

d)   All registered documents should reflect approved changes resulting from change orders. Details of the approved changes, such as date, request number and description, should be recorded in the relevant documents, and the updated documents should be transmitted to the appropriate project team members.

5.10 SUMMARY

The project manager and construction manager must ensure proper and sufficient laydown areas for the pipe spools to be stored are provided with easy access and retrieval. Storing the pipe spools according to the construction sequences is yet another brilliant idea, which is mostly ignored at laydown areas. Often in laydown areas, piping materials are laid down without proper identification and grouping, which leads to many complications later, and must be prevented. Proper inventory management of the piping material stock is another critical point to look at. One more important area to be addressed is receiving the right type of valves with proper documentation. Many issues have also surfaced in this area. Handling, preservation and issue of valves must be taken care of. The QA/QC manager must ensure that all the materials received at work site for erection are supplemented with proper documentation. Where possible, pre-inspection of items at source with client will reduce the risk of rejection. The most common phenomena in piping project sites are site modification and additional piping works requested by the client. The construction manager needs to ensure there are sufficient people to address this area. Proper documentation related to site modification and additional piping requests needs to be ensured to bill the client accordingly.

Section 6

Pre-commissioning

After reading this section, you should be able to understand the importance of . . .

1. Pre-commissioning line walk and punch listing of defects.
2. NDT clearance, punch list closing, test packs and ITRs management and approvals by client.
3. Regulatory approvals for pre-commissioning.
4. Releasing the piping system for pressure testing, dewatering, reinstatement.
5. Bolt tightening and flange management.
6. Preservation, paint touch-up, insulation and line numbering.
7. Mechanical acceptance by client.

PRE-COMMISSIONING
↓
Line Walk Planning
↓
Punch List Identification
↓
Test Packs / ITRs Control
↓
NDT Clearance
↓
Punch List Closing
↓
Test Packs Approval
↓
Release for Pressure Test
↓
Regulatory Approval
↓
Pressure Testing & De-Watering
↓
Reinstatement and Preservation
↓
Line Numbering
↓
Bolt Tightening & Flange Management
↓
Paint Touch-Up and Insulation
↓
MC Acceptance by Client

6.0 INTRODUCTION

Upon completion of the pipe spools erection, the next important activity is to do the line walk with the client to identify the integrity of the piping structure in the systems. Deviations must be recorded in the form of punch lists and necessary corrective/preventive actions shall be carried out. In most projects, there are many punch list defects, which is a clear indication of the poor quality of construction work at site. This needs to be taken care. Good-quality construction work will lead to fewer punch lists at the line walks. As part of the pre-commissioning works in the oil and gas projects, all the inspection and test reports (ITRs), test packs and the related documentation need to be properly completed and signed off by the client. This is yet another area where the contractors face difficulties in managing, thus leading to delays in completing the punch lists and outstanding documentation. Regulatory approval for oil and gas piping systems pressure testing is an important activity that needs to be completed on time to avoid delays. As safety of the plant and the people involved in the piping system pre-commissioning works is paramount, necessary care and planning for regulatory approvals should be taken at appropriate times. In this section, most of the pre-commissioning related essentials are discussed in detail.

6.1 LINE WALK

It is important to make sure that all pipes are properly installed and functioning as per design. Improper installation or malfunctioning of even a single support may upset the whole piping system and create unbalanced forces throughout. Therefore, it is critical to do a line walk of the whole piping system and inspect each and every pipe support on the line, even though the overall design may incorporate several different types of pipe supports.

Piping systems consist of several components to keep the piping properly suspended and installed. The deflection and movement during installation is compensated by different types of pipe supports, which keep the piping system balanced. In the majority of cases, the pipeline runs between two pieces of fixed equipment. The fixed equipment components act as the anchor points of the pipeline. Pipe supports are located between these anchor points and provide mobility throughout the piping system.

To understand the critical nature of properly functioning pipe supports and how they contribute to the overall system, one should thoroughly inspect each component for proper operations during different phases of the life of the unit. This inspection survey could be done in the following phases:

Line walk performed as part of pre-commissioning criteria should focus on the following:

a) The actual operating conditions do not exceed anticipated operating conditions, including loading and/or movement of the pipe supports.
b) Pipe supports, or any components thereof, have not been damaged during fabrication.
c) Check for any discrepancies from actual installation and drawing.
d) Perform redline markup as required.
e) Capture all findings into final documentation.
f) Record punch list.

Documents required:

a) Piping isometric drawings
b) Facility general arrangement drawings

6.2 PUNCH LIST

The punch list items identified during line walk, ITR-A, ITR-B and ITR-C should be rectified on the spot or entered into the database.

A punch list item is work that is incomplete or not installed or not tested in accordance with design or codes and specification. A punch list is raised, if required, when a system has been handed over as mechanically completed or commissioned and a number of such items is outstanding.

Punch list items are categorized as shown below:

CAT A: Items that prevent the safe operations of plant and equipment in its final operation mode and would present a safety hazard to personnel operating the equipment. "A" items are classed as major punch items.

CAT B: Items that cover work that is incomplete or not in accordance with design but that does not prevent commissioning or operating activity or cause a safety hazard. "B" items are classed as minor punch items. Items may be completed after acceptance or handover subject to approval by relevant parties.

For reporting of the activities, refer to the following sample formats:

Sample Reporting Formats		
Annexure – 29	:	Punch List (Form No. PCR-QAF-29)

6.3 TEST PACK/ITRS (INSPECTION AND TEST REPORTS)

ITRs shall be completed during construction and checked against every discipline, which will confirm correct and complete installation. The construction mechanical completion team is responsible for performing the ITR-A/B. The minimum signatories for ITR-A/B are the discipline engineer or project engineer, and review is by the commissioning manager/HUC coordinator.

ITR No.	Construction	ITR No.	Pre-Commissioning	ITR No.	Commissioning
P01-A	Pipe Work Completion	P01-B	N/A	P01-C	N/A
P02-A	Pre-Test	P02-B	N/A	P02-C	N/A
P03-A	Flushing	P03-B	N/A	P03-C	N/A
P04-A	Pressure Test	P04-B	Reinstatement Work	P04-C	System Commissioning Leak Test
P05-A	Blowing/Drying	P05-B	N/A	P05-C	N/A
P06-A	Flanges Alignment	P06-B	Flanges Management	P06-C	N/A
P07-A	Bolt Tensioning/ Torquing	P07-B	N/A	P07-C	N/A

ITR No.	Construction	ITR No.	Pre-Commissioning	ITR No.	Commissioning
P08-A	Chemical Cleaning	P08-B	N/A	P08-C	N/A
P09-A	Re-Installation Work	P09-B	N/A	P09-C	N/A
P10-A	Service Test	P10-B	Service Test	P10-C	N/A
P11-A	Pipe Painting/ Coating/Marking	P11-B	N/A	P11-C	N/A
P12-A	Pipe Insulation	P12-B	N/A	P12-C	N/A
P13-A	Critical Valve Leak Test	P13-B	N/A	P13-C	N/A
P14-A	N/A	P14-B	Witness Joint	P14-C	N/A
P15-A	N/A	P15-B	Lube Oil Flushing	P15-C	N/A
P16-A	N/A	P16-B	N/A	P16-C	Nitrogen Purging

6.3.1 Piping testing and pre-commissioning common scope of work

a) Verify conformity line checking against IFC process and instrumentation diagram (P&ID) drawings as follows:

  i. Verify valves, flanges, blind flanges and piping items are properly installed and tightened.
  ii. Verify gaskets, bolts and nuts installed are of the right specification.
  iii. Verify the flow direction of check valves, globe valves, restriction orifices, and other in-line instruments.
  iv. Verify pipe works are adequately supported as per drawings.
  v. Verify the accessibility and operability of manual valves and all instrument indicators.
  vi. Verify all process pipes, process equipment, line instrument and isolation valves have been leak tested.
  vii. Verify color coding, line numbering and directional arrow are provided correctly.
  viii. Verify functional test of all manually operated, isolation and block valves.

b) Verify flushing has been completed, which shall be referred to flushing/hydro testing procedure for detailed flushing procedure.

c) Verify hydro testing has been completed, which shall include the following activities:

  i. FAT/SAT test of valves has been done based on acceptance.
  ii. Criteria on API 6D and BS 5351 Codes and Standards.
  iii. Non-return valves and swing check valves shall be returned to their original position after completion of hydro testing.
  iv. Temporary spade and blinds are removed after hydro testing.
  v. Pipe works are drained and blown with dry air once hydro testing is completed.
  vi. All instruments and relief valves that are removed during testing are reinstated to their original location.

d) Perform reinstatement leak test, which shall include the following activities:

    i. Perform internal inspection and verify cleaning of vessels is completed and "Final Closure Certificates" are issued.

    ii. The flange management shall be performed prior to reinstatement leak test, which consists of:

        1. Flanges alignment to ensure the correctness of flange-to-flange alignment, gaskets and cleanliness of flange faces.

        2. Bolt tensioning/torquing as per approved procedure to ensure the tightness of stud bolts with correct tension/torque value, including the correctness of stud bolt installation.

        3. Gross leak test with seven bar compressed air shall be performed prior to reinstatement leak test to ensure the early detection of the major leaking points and then to avoid any wasting of nitrogen consumable.

        4. Verify test instruments used have a range of 50% higher than the test pressure.

        5. Upon completion of reinstatement leak test, the system shall be preserved by one bar nitrogen.

e) Verify insulation and claddings conform to specification and drawings.

f) Verify pipe work is internally cleaned and preservation applied after pickling or acid cleaning.

g) Verify the systems have been purged with nitrogen and achieve 5% oxygen content.

h) Verify the baseline survey (BLS) on the basis of corrosion risk assessment for corrosion management of piping system.

6.4 NDT CLEARANCE

Inspection shall be carried out after a welded component is completed:

a) Weld size and reinforcement
b) Completeness of all welds
c) Appearance of welds
d) Geometrical and surface defects on welds
e) Identification of welds and welder numbers
f) Distortion of components
g) Arc strikes on parent metal
h) Dimension check
i) Straightness check
j) NDE inspection

The following records shall be produced:

a) Fit-up report
b) NDE report
c) Weld visual report
d) Lamination report
e) Dimensional report

6.5 PUNCH LIST CLOSING

The punch list items identified during line walk, ITR-A, ITR-B and ITR-C shall be closed and witnessed by relevant parties by signing off the punch list form.

A punch list item is work that is incomplete or not installed or not tested in accordance with design or codes and specification. The punch lists are raised, if required, when a system has been handed over as mechanically completed or commissioned and a number of such items is outstanding.

Punch list items are categorized as shown below:

CAT A: Items that prevent the safe operation of plant and equipment in its final operation mode and would present a safety hazard to personnel operating the equipment. "A" items are classed as major punch items.

CAT B: Items that cover work that is incomplete or not in accordance with design but that does not prevent commissioning or operating activity or cause a safety hazard. "B" items are classed as minor punch items. Item may be completed after acceptance or handover subject to approval by relevant parties.

For reporting of the activities, refer to the following sample formats:

Sample Reporting Formats		
Annexure – 29	:	Punch List (Form No. PCR-QAF-29)
Annexure – 30	:	Punch List Tracking Summary (Form No. PCR-QAF-30)

6.6 TEST PACK/APPROVAL

Each test pack shall contain the following as a minimum requirement:

a) **Test pack criteria**

The test criteria shall define the types, limits and duration of the test. The criteria shall correspond to the equipment specifications and aim to demonstrate proper operations and performance of that equipment. The commissioning team shall review equipment specifications and vendor's operating manuals to develop specific testing criteria for equipment according to which sub-system it belongs.

b) **Test pack preparation**

The test preparation shall outline the associated supporting utility and protective equipment that are required to be commissioned and serviced prior to testing of the subject equipment.

c) **Test pack procedure**

The test pack procedure shall describe in detail the following:

i. *Description of the test*

A description of the test that is to be performed during the testing and pre-commissioning activities, including as a minimum:

1. The type of function to be verified
2. The type/size of load or stimulation to be applied
3. The duration of the test
4. The test parameters to be monitored and recorded
5. The mode of operations
6. Test line arrangement

The test line arrangement shall indicate the route of the test fluid or power. The type/source of fuel or power supply is used for testing. Any temporary piping or electrical wiring required for the test must be specified. Marked-up P&ID or sketches showing the test line arrangement and isolation points must be attached with the procedure.

ii. Sequence of the tests

A logical sequence of the tests shall be presented. Generally, the functional tests for the control shall be performed prior to the load tests.

iii. Test data and records

Test and PC record sheets shall be generated together with the procedure. Equipment and instrument tag number, manufacturer, model and serial numbers, capacity, normal/design operating parameters, set points and tolerances in addition to test data shall be included in the test record.

A complete set of drawings shall be maintained during equipment installation/assembly. All changes made during installation, testing and commissioning shall be reflected immediately on the redline as-built drawings.

d) Specific responsibilities

The test pack is written by the commissioning engineer and approved by the commissioning manager. On completion, the following signatures are required:

i. Commissioning engineer
ii. Commissioning manager

For reporting of the activities, the following sample formats shall be used:

Sample Reporting Formats

Annexure – 31　　　　　　:　　　　　　Test Packs Reporting Summary (Form No. PCR-QAF-31)

6.7 RELEASE FOR PRESSURE TEST

Prior to release for pressure test, the following shall be fulfilled and signed off by the commissioning engineer and manager.

a) **Pipe work completion**

   i.   Check the piping is constructed according to the latest IFC revision of P&ID and isometric drawing.
   ii.   Check that materials used are as per specification.
   iii.   Check that piping routing does not obstruct the access for emergency escape and maintenance.
   iv.   Check piping line sloping, if required.
   v.   Ensure maintenance accessibility for valves, PSV, sample point, corrosion probe, etc.
   vi.   Check that flange rating or gray lock hubs are correct.
   vii.   Check that gaskets are provided as per specification and QA/QC report.
   viii. Check that flange joint is properly aligned and within acceptable tolerance and tallies with QA/QC report.
   ix.   Check bolts and nuts are adequately tightened and of correct length.
   x.   Check that required vent and drain points are provided.
   xi.   Inspect all valves and ensure that the type, size, material and flow directions are correct.
   xii.   Check filter/strainer body and inspect internal elements for damages.
   xiii. Check that filter/strainer element materials and mesh size are correct.
   xiv.   Check that pipe supports are constructed and installed according to drawings.
   xv.   Check that welded joints have been completely inspected.
   xvi.   Check that actuated valves are correctly installed.
   xvii. Check that piping penetrations are constructed according to drawings with sufficient stiffening plates and adequate sealing.
   xviii. Check that all piping special items are installed properly.
   xix.   Check that spectacle blinds, spades and spacers are provided and installed at location and positioned as per P&ID.
   xx.   Ensure exposed portions of spectacle blinds are coated with grease to avoid corrosion.
   xxi.   Ensure that all instrument drains lead to the drain system.
   xxii. Verify that check valves and globe valves are installed in their correct flow direction.
   xxiii. Check that lines are labelled at visible locations.
   xxiv. Check that mating of dissimilar material is installed correctly.
   xxv.   Check that all chain wheels and extended spindles required for specified valves have been installed.
   xxvi. Check the isolation gasket is installed, if required.
   xxvii. Ensure PSV system is installed and calibrated.
   xxviii. Check the spring hanger condition is in locked position, if required.

b) **Pre-hydro test**

   i.   Verify to ensure that all piping construction/installation activities as per test package limit have been completed.
   ii.   Ensure that the test package can be applied.
   iii.   Check that all weld inspection and NDE are complete with traceability and history records complete and cleared by QA/QC department.

    iv. Ensure that conformity line check has been performed.

    v. Ensure that test package, including the attachment, is completed.

    vi. Check that all vents and drains are installed (add temporary vents and drains for testing as required).

    vii. Ensure all check valves are removed or tied back.

    viii. Ensure ball valves are half-opened during test and fully opened for flushing.

    ix. Verify that globe valves are fully opened for testing and flushing.

    x. Check that all temporary equipment stops, spades, bobbin pieces and supports for pressure test are installed and recorded for checking at post-test reinstatement.

    xi. Check that all instrument relief valves, bursting discs and specified control valves have been removed for pressure test.

    xii. Ensure all orifice flanges are blinded off.

    xiii. Ensure essential supports are fitted and springs and bellows are gagged during test.

    xiv. Check that all sprinkler heads, rosettes and nozzles have been removed and connections plugged.

    xv. Raise piping punch list and ensure all Category A items are cleared.

c) Flushing

    i. Verify that all piping construction/installation activities as per test package limit have been completed.

    ii. Check that injection points and drain-out points are completed as per test pack markup.

    iii. Ensure that water quality is good, with maximum 30 ppm chlorite and demineralized water for stainless steel line.

    iv. Verify equipment is bypassed and blinded.

    v. Check valve status in accordance with flushing and testing procedures.

    vi. Ensure all low point drains and high point vents are flushed.

    vii. Ensure that all piping lines in this test package have been flushed.

    viii. Ensure that piping system has been drained out, followed by drying.

    ix. Protect/preserve the piping system. If it cannot be done immediately, ensure that hydro testing of the same piping is completed.

    x. Record data as below:

Injection Point	Flow Rate	Drain Point	Result

6.8 REGULATORY APPROVAL

The PCR needs to ensure that all the regulatory approvals related to the piping works carried out in the project are obtained on time. Piping works, which involve hydro testing, must be witnessed by DOSH for safety compliance during testing, and BOMBA must be witness to all the fire water pipes for fire water pressure testing. DOSH approvals are also required for PMT equipment number certifications connected to the piping.

Care should be taken to ensure that both client's and regulatory authorities' approval is obtained from the testing stage to obtain a certificate of fitness (CF) once the system is ready.

It is better to inform the authorities well in advance of the testing plans, so that the officers in the regulatory bodies are available in a timely manner for inspection and testing.

If authority approvals are part of the scope of the client, then necessary follow-up and planning should be established in discussion with the client.

Regulatory approvals as part of the subcontractor's scope of works must be planned accordingly.

The PCR must ensure that a copy of all the regulatory approval records is maintained for future reference and for mitigation of contractual issues, if any.

6.9 PRESSURE TESTING

a) **Hydro test**

i. Verify that all piping construction/installation activities as per test package limit have been completed.
ii. Ensure that water flushing has been completed.
iii. Ensure that pre-test preparation is complete and pre-hydro test is attached together with P&ID where test limits are clearly marked.
iv. Ensure preliminary piping punch list is attached and all "A" items on line are cleared and signed off.
v. Ensure calibration certificates for recorders and gauges are attached.
vi. Ensure instruments as stated in CSP 22 are removed.
vii. Ensure test temporaries are installed.
viii. Ensure permit to work or HSE authorization is in place.
ix. Verify release for pressure test.
x. Record data as below as per CSP 22:

Pressure	Temperature	Remark
0%		
25%		
50%		
75%		
100%		

xi. Pressure test to complete.

6.10 DE-WATERING

a) **Blowing/drying**

i. Verify to ensure that pressure test has been completed.
ii. Ensure that water flushing has been completed.

iii. Ensure that the test medium has been drained upon completion of the hydro test.
iv. Ensure that no water is trapped in the system.
v. Ensure that blowing medium/compressed air humidity is $-30°C$.
vi. Ensure that all the blowing point humidity is $-10°C$.
vii. Ensure that permit to work or HSE authorization is in place.
viii. Record data as below:

Blowing Injection	Blowing Point	Humidity	Result

6.11 REINSTATEMENT

1. Flushing and hydrostatic testing have been satisfactorily completed and recorded. Check that drying of lines is complete. Verify that all in-line units have been reinstated as per the latest P&ID.
2. Ensure correct bolts and gaskets are installed at all mechanical joints, and bolts are torqued/tensioned to correct specification.
3. Check all manual valves are in open position (recommended) and smoothly turned.
4. Ensure all temporary spades, bobbins, vents and drains have been removed and openings closed according to specifications.
5. Ensure support gauges are removed where applicable (leaving some in place for removal offshore may be required).
6. Ensure all process blinds, spades, valves and in-line equipment have been correctly installed as per P&ID.
7. Check/ensure that lines are resting firmly/properly on their supports (not floating).
8. Check that extended spindles and chain wheels for specified valves have been correctly installed.
9. Check that all locking devices and interlocks for specified valves have been correctly installed.
10. Ensure piping, instruments, mechanical equipment and specialty items are reinstated in accordance with the P&ID. Verify the items against pre-test list.
11. Verify earthing strap/connection, if required.
12. Verify the baseline survey for corrosion management is completed.

6.12 PRESERVATION

Preservation is carried out to maintain the pre-commissioning conditions of the installation until commissioning comes into force.

a) Preservation with dry air

If the initial gas filling is carried out later than six months after the installation works are completed, preservation of the piping takes place according to the dry air method. The main criteria for a dry pipeline are as follows:

i. Dew point temperature must be $\leq -20°C$ measured at atmospheric pressure.

ii. Overpressure after sealing of the pipelines must be at least one bar.
iii. The acceptable minimum norm during the entire preservation period is a measured dew point temperature of $\leq -10°C$ (measured at atmospheric pressure and at an overpressure of one bar).

The dew point temperature and the pressure are measured and registered within the preservation period with stated intervals since the contractor as well as a neutral third party is involved.

b) Preservation with nitrogen (N_2)

If preservation is decided by using nitrogen, the criteria are as follows:

i.   Purge with nitrogen until achieving 100% N_2.
ii. The overpressure after sealing must be at least two bars.

Measurement of the pressure at stated intervals during the preservation period is required.

6.13 LINE NUMBERING

Piping lines shall be numbered in the following manner:

a) Numbering of all lines excluding steam tracing spools

i. Piping serial number, in general, is started from 0001 onwards except for the units that are characterized by more than one section, such as crude and vacuum distillation units. In such cases, the splitting of piping serial numbers to be assigned for each section of the unit shall be determined by the contractor. Special number 7001 : 9999 shall be used for all drains, relief headers and utility services, including fuel oil and fuel gas for all units, except for the units that are producing the subject utility services. For assigning the piping serial number, the following items should be taken into consideration:

1. Pipe line numbers shall be prefixed, from source to unit battery limit, with the unit number of the unit of origin.
2. The individual line number shall be held up to the point where the line ends at the inlet of equipment, such as a vessel, exchanger, pump, etc. Another number is required for the line downstream of the equipment.
3. A new line number is required when the pipe design condition can vary (e.g., downstream of the control valve assembly) or when a new piping class is to be specified.
4. Line number shall be held up to the point where the line ends to the header or unit battery limit block valve. All branches to and from header shall have an individual line number.
5. All utility headers (systems) shall be numbered with their respective units. All branches serving a specific unit shall be numbered with that unit.
6. All fire water and sewer branches serving a specific unit shall be numbered in accordance with the note above.

ii. Piping class code shall be in accordance with the line classes utilized in project piping material specification.

iii. Piping components not identified by instrument or mechanical equipment numbers, etc., and not covered by the piping material specification, are identified by a special item number.

iv. Unit number of the plant shall start from 1 (not from 01).

b) **For steam tracing spools for steam tracing numbering and material takeoff, the contractor can use their own system.**

Example:

3"-P-12007-A11A-H30

3"	–	Signifies the line sizes in inches
P	–	Signifies fluid service
12007	–	12 – signifies unit or facility number 007 – denotes the serial number
A11A	–	Denotes the piping service class A – denotes the flange rating 11 – denotes the piping material A – a suffix qualifying the piping material
H	–	Denotes the insulation type
30	–	Denotes the insulation thickness

6.14 BOLT TIGHTENING AND FLANGE MANAGEMENT

a) **Bolt tightening**

i. Check stud bolts and nuts for diameter, type of threads, material, coating and condition in line with isometric drawings.

ii. Lubricate the stud bolt threads and nut faces.

iii. Check that nuts are fitted correctly.

iv. Tighten sequentially according to procedure and the required torque.

v. Ensure that stud bolt length is sufficient as per standards.

vi. Check lock nuts are required.

b) **Flange management**

i. Verify the check sheets for alignment of flanges and proper tightening of bolts and ensure that they have been completed and approved.

ii. Ensure correct bolts and gaskets are installed at all flanges joints and bolts are torqued/tensioned to correct specification and certified.

iii. Check all the flanges have been tagged and identified in the database to ensure that the following items are corrected:

  • All flanges data with flange connection number
  • Drawing number
  • Isometric line number
  • Sub-system number

- Alignment check (date and signature)
- Cleanliness check after hydro test or flushing (date and signature)
- Gasket type and rating
- Bolt type, diameter and length
- Brand of lubricant applied
- Used tool (number and type)
- Torque value or tensioning pressure (when applicable)
- Date of execution
- Name of the operator

6.15 PAINT TOUCH-UP

a) Verify that all piping activities have been completed.
b) Ensure that all piping ITRs have been completed and are available.
c) Ensure sharp edges are removed prior to application.
d) Ensure all masking is completed.
e) Ensure piping supports are painted.
f) Ensure spot painting is performed.
g) Ensure dry film thickness is checked and corrected.
h) Record any defective area(s) on punch list.
i) Check attached drawing(s) with highlighted piping sections.
j) Ensure line numbering is according to P&ID.
k) Ensure flow direction is according to P&ID.
l) Ensure color coding is according to project requirements.
m) Ensure tagging of manually operated valves is correct.
n) Check that all safety/warning signs are in correct locations.

Note: This painting/coating is also applicable to
SS/DSS/SDSS piping as per project paint specification.

6.16 INSULATION

a) Check that insulation installation is in accordance with specification and AFC drawings.
b) Check that insulation sealing is completed to prevent the ingress of water or moisture.
c) Check that access point is provided in accordance with specification for corrosion monitoring.
d) Ensure that insulation is installed after NDT hydro test and painting is completed. Perform paint touch-up, if required.
e) Check that the direction of flow/color coding is indicated on the insulation cladding.
f) Ensure that oil, grease, rust and other foreign matter are removed from the piping outer surface prior to installation of insulation.
g) Check that longitudinal joints between segments are staggered.
h) Ensure that the insulation system is provided with 15 mm diameter drain hole for every three meters length and/or follows the approved project insulation specifications.
i) Check that personal protection items, such as open guard, shield, wire mesh or railings do not obstruct operational access.

j) Ensure that removable insulation for flanges and valves shall be suitable for quick removal and reinstallation.

k) Produce a defect list for issuance of punch list.

6.17 MECHANICAL ACCEPTANCE BY CLIENT

In order to obtain mechanical completion certificate, ITR-A and ITR-B shall be completed and punch list Category A shall be cleared. All documents shall be compiled and signed off by relevant parties, i.e., commissioning engineer, commissioning manager and client.

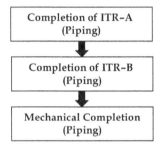

The mechanical certificate shall be signed off by the client's project manager as shown in the sample below.

MECHANICAL COMPLETION CERTIFICATE

Contract No :
Contract Title:
Contractor   :

In respect of the above CONTRACT, COMPANY hereby issues to the above-named CONTRACTOR this Mechanical Completion Certificate.

This Certificate shall not relieve CONTRACTOR of its continuing obligations to COMPANY under the CONTRACT, nor shall it affect statutory or common law rights held by COMPANY.

This Mechanical Completion Certificate is issued acknowledging that WORK has been completed checking for the following Punch List Items and COMPANY-deferred WORK.

Refer to attachment for offshore hookup workscope and punch list.

Mechanical Completion Date:
Issued for COMPANY:
By   :
Name :
Title :
Date   :

6.18 SUMMARY

As the construction activities in the project progress, the construction manager and the QA/QC manager together with the client should ensure that a proper pre-commissioning plan is established and rolled out in phases. Complete coordination between the construction and QA/QC teams are essential at this point of the project as this will lead to a thorough examination of the work at the site for quality and technical specification conformance together with the required documentation. Construction engineers and QA/QC engineers of the respective disciplines must take proactive roles to ensure that their respective discipline line walk, punch list identification, closing, ITRs, test pack and mechanical acceptance by the client are established. The challenges in pre-commissioning are: poor documentation, tracking of MC line walk punch lists, ITRs and test packs compilation. The QA/QC manager must proactively plan for necessary resources to ensure pre-commissioning activities are carried out smoothly. For some areas, regulatory approvals are required by agencies, such as the DOE, DOSH, Majlis, etc. This needs to be clearly planned and communicated well in advance to avoid last-minute cancellations and postponement, which can be costly.

Commissioning

After reading this section, you should be able to understand the importance of . . .

1. Mechanical completion (MC) line walk, punch list identification and closing.
2. MC documentation sign-off and MC certification from client.
3. Release for commissioning.
4. Site modification works.
5. Project final documentation works and handover to client.

© Can Stock Photo

COMMISSIONING
↓
MC Line Walk
↓
MC Punch List Identification
↓
Punch List Closing
↓
MC Documentation
↓
Mechanical Completion Certificate from Client
↓
Release for Commissioning
↓
Site Modification
↓
Project Final Documentation

7.0 INTRODUCTION

Commissioning is a crucial part of the project and is the final stage of the piping project completion. A carefully planned sequential line walk, punch list identification, closing of punch list and regulatory approvals will lead to smooth commissioning of the project. Similarly, the final documentation of the piping system-related works should be carried out to ensure appropriate sign-off of commissioning by contractor, client and regulatory officials. Successful commissioning is a clear indication of the project's success. In many projects, during the commissioning stage, accidents and incidents have happened, which is a costly affair. Prior to commissioning, all the necessary protocols must be adhered to, to prevent unexpected results. If commissioning falls under the client's responsibility, then the contractor is at ease. However, quality of piping works results appropriately in the commissioning. In most oil and gas projects, one of the issues faced by the PCR is the timely completion of final documentation, which leads the contractor to spend a good amount of time and resources on it. A properly planned final documentation team can help to prevent this issue. Many of the clients keep 5% to 10% of the project sum as the final documentation retention, which is a large sum for the contractor but can be easily obtained if project final documentation is completed and handed over on time. In this section, some of the above topics and their critical DOs and DON'Ts are discussed in detail.

7.1 MECHANICAL COMPLETION (MC) OF LINE WALK

Piping line walk before commissioning shall revisit the ITR-A, ITR-B and punch list during pre-commissioning. It must be ascertained that all test packs and findings have been closed out.

7.2 MC PUNCH LIST

All Category B punch lists captured during line walk prior to commissioning work shall be rectified and cleared. It shall be witnessed and signed off by the client.

For reporting of the activities, refer to the following sample formats:

Sample Reporting Formats		
Annexure – 29	:	Punch List (Form No. PCR-QAF-29)

7.3 PUNCH LIST CLOSING

Prior to pre-commissioning and commissioning, the contractor shall carry out line check/ line walk to check the facilities for correct erection and installation, operability, maintenance requirement and safety of plant and personnel during operations in accordance with accepted engineering practices. The company representatives must be present to witness this. The contractor shall carry out all modifications and corrections in accordance with the final punch list prepared jointly by the contractor and the company.

7.4 MC DOCUMENTATION

ITR-A and ITR-B with supporting documents, such as mill certificates, calibration certificates, etc., shall be made available prior to commencing ITR-C. Below is the ITR-C for piping commissioning:

a) Reinstatement leak test

   i.   The leak test has been completed and accepted as per company procedure.

   ii.   Leak test result (leak rate and maximum permitted leak rate).

   iii.   The completed system has been de-pressurized and all signs and barriers have been removed.

   iv.   All test equipment has been removed and line reinstated using witness joint procedure.

   v.   All associated P&IDs have been redlined as-built.

   vi.   The completed system shall be preserved with nitrogen (N_2) gas at one bar of pressure.

b) Witness joint

   i.   Prior to conducting reinstatement leak test, flange faces have been examined and seen to be clean and damage free.

   ii.   Gaskets are of correct type and size.

   iii.   Nuts and stud bolts examined and seen to be not damaged and of correct size and type.

   iv.   Flange faces parallel.

   v.   Joint pulled up evenly on diametrically opposing stud bolts.

   vi.   All nuts pulled up to required tightness.

   vii.   No visible damage to gasket when joints completed.

   viii. Maximum allowable of two threads through each nut.

   ix.   No excessive length of stud bolt protruding through each nut.

c) Purging

   i.   Verify that the reinstatement leak test for the system/sub-system/test package above has been completed and accepted.

   ii.   The system shown on the attached P&ID (test package) has been purged and achieved 7% O_2 content.

7.5 MECHANICAL COMPLETION CERTIFICATE

Mechanical completion of a system includes the following but is not limited to:

a)   All design and engineering work has been completed.

b)   All installation work for that system, including equipment, has been completed in accordance with "Approved for Construction" drawings, specifications, applicable codes and regulations and good engineering practices. All tie-in connections have been made, all testing and inspection completed and system/facilities are ready for pre-commissioning.

c)   All instruments have been installed.

d) All factory acceptance tests and other testing and inspection activities have been completed.
e) The contractor has obtained all relevant approvals from the company.
f) Safety studies have been completed and requirements of all the safety studies have been met and all documentation put in place.
g) All required documentation and certification documents required by the contract have been supplied.
h) All operating procedures and maintenance procedures have been forwarded to the company well in advance for review.
i) All items for which the contractor is responsible for obtaining third-party, regulatory or company approval have been obtained, and confirmation documentation has been provided to the company.

Mechanical completion is defined as the state where all systems, including utility and auxiliary facilities, have reached the condition of pre-commissioning stage and accordingly been certified by the company.

Contract Number: CHO/2013/TOG/0001

Notice of Mechanical Completion

It is hereby certified that the scope of work referenced has been inspected and/or tested and is Mechanical Completed, in accordance with the Contract requirements, and Contractor is permitted to proceed with Pre-commissioning and Commissioning.

Contractor		Date	
Name		Position	Construction Manager
Contractor		Date	
Name		Position	Project Manager
Owner		Date	
Name		Position	Project Manager

7.6 RELEASE FOR COMMISSIONING

"Release for commissioning" means the point at which the company issues a certificate to the contractor saying that the system is ready for commissioning. Ready for commissioning status shall be jointly reviewed by the company/contractor.

At this point, all systems and equipment shall be at a stage where process fluids can be safely introduced and equipment can be safely operated with all control and safety devices in-service/functional to meet the requirements as per design and specifications.

General

1. Are all the required utility support services for this activity available and adequate? For example, water, electricity, cool air, nitrogen, etc.
2. Have all assigned engineering and construction punch lists and review issues prior to this activity been addressed?
3. Has construction and equipment been checked to ensure that it is consistent with the P&IDs and all design specifications?
4. Have all action items for design review, HAZOP and concept review been resolved?
5. Have utility systems been isolated from process systems? Are flanges or blinds inserted, check valves closed, circuit breakers opened?
6. Have all the necessary manpower requirements been addressed? For example, organization chart (with names and contact numbers).

Electrical/instrument

7. Have all computer control logic and software been thoroughly tested to ensure that they are working as intended?
8. Has instrument grounding been coordinated with cathodic protection for pipes, tanks, structures, etc.?
9. Are all electrical grounds intact?
10. Is all equipment compatible with the area's electrical classification?
11. Have all motors, feeders and transformers installed been properly inspected?
12. Has all flow-regulating equipment, such as valves, louvers, dampers, etc., been tested?
13. Has all relevant electrical testing, e.g., continuity test, functional test, etc., been completed and approved by the relevant authority?
14. Has all relevant instrument testing, e.g., megger test, continuity test, etc., been completed and approved by the relevant authority?

Engineering/construction

15. Have mechanical safety systems, including relief valves and ventilation, been tested?
16. Are there sufficient devices (drain bleeders, vent bleeders, etc.) to properly prepare equipment for this activity?
17. Has rotating equipment been checked for rotation, alignment and lubrication?
18. Have piping, tanks and vessels been properly hydro tested or tested per an alternate method?

19. Have pipe supports been installed so that no piping strain exists?
20. Has the use of SBF been minimized? Are all SBFs adequately protected from excess vibration, etc., which can cause leaks?
21. Have the correct bolting (no short studs) and proper gasket materials been installed?
22. Are all corrosion protection systems in place, such as corrosion inhibitors, cathodic protection, coatings and paintings?
23. Have all screwed fittings in hydrocarbon service seal been welded as required by PTS?
24. Have all construction blinds been pulled?
25. Have steps, such as tightness and leak tests, for example, operating pressure leak test or torquing, been completed?
26. Have all relevant mechanical testing, for example, test run, etc., been completed and approved by the relevant authority?
27. Has all relevant civil testing, for example, concrete cube test, etc., been completed and approved by the relevant authority?
28. Have all relevant non-destructive tests, i.e., RT, DPT, PWHT, PMI, MPI and UT been completed and signed off by the relevant authority?
29. Has the inspection team been notified of this activity, so that they can notify the appropriate government agencies?

Human factors/ergonomics

30. Has training of each employee involved in this activity been identified and completed?
31. Is there a need for additional access/work platforms for this activity?
32. Is access to equipment and valves adequate for this activity?
33. Is there adequate lighting at the equipment/work area?
34. Is there adequate lighting on the equipment (such as sight glasses, etc.)?

Process safety information (PSI)

35. Have all relevant PSIs been marked up/updated (PEFS, electrical, instrument, structure, mechanical, etc.) as appropriate?
36. Has it been ensured (i.e., checked) that the documents for safety, process, equipment, instrumentation, etc., have been updated? For example, safeguarding memoranda, operating procedures, equipment data sheets, mechanical and instrumented safety system documents, etc.
37. Are the PEFSs marked up for submission correctly drawn, including the notes section?
38. Has all relevant process safety information been reviewed and updated?

Health, safety, environment and emergency response

39. Are adequate signs/barricades/safety measures available and in the correct places?
40. Are all potential releases in the process area adequately contained?
41. Are there suitable barricades between process equipment and adjacent roadways?
42. Is suitable PPE for this activity available and adequate?
43. Is there adequate access to the equipment isolation devices during this activity and in emergencies?

44. Have all inherent dangers in the immediate vicinity of the equipment isolation devices (hot piping, sharp points, etc.) been removed?
45. Is there sufficient hazardous energy isolation?
46. Are overhead pipe racks protected from crane/lifting equipment impacts?
47. Have all non-essential vehicles/equipment been removed from the process area?
48. Has the safety group been notified of this activity, so that they can notify the relevant government agencies?
49. Have all PTWs prior to the activity been duly closed?
50. Has the environmental team been notified of this activity, so that they can notify the relevant government agencies?
51. Are environmental control devices in good order?
52. Have construction waste/materials been disposed of in an acceptable manner? Is the area clear to allow this activity?
53. Is all necessary firefighting equipment available and easily accessible for this activity?

Other issues/concerns

54. Has adequate notice been given to people working in the surrounding area?
55. Other issues?
56. Other activities are planned that coincide with the commissioning.
57. Insulation.
58. PI Tag creation.
59. Baseline reading for vessel thickness.
60. Master data list.
61. Commissioning procedure.

7.7 SITE MODIFICATION

Carry out site modifications as found necessary during system check/inspection (line check) from the viewpoint of routine operation, maintenance and safety of the OSC plant. A list of such jobs shall be prepared by the company and shall be handed over to the contractor for execution.

7.8 FINAL DOCUMENTATION

- Final documentation (also known as turnover documents) for the project is very critical for oil and gas projects. This is especially so for piping.
- Firstly, project-specific final documentation procedure must be prepared by the PCR and reviewed and approved by the client.
- Final documentation procedure of the project must be in line with the ITB or tender/contract documents.
- If clients insist on using their final documentation procedure for the project, it is well and good. However, care should be taken to ensure excessive documentation-related works and costs are not involved.
- The PCR must establish a final documentation team depending on the size and complexity of the project and the amount of piping works involved.

- A final document controller (also known as lead DC or lead turnover engineer) shall be identified and appointed from the beginning of the project to prevent missing of critical information related to the project as well as to avoid delays in compiling the final documents at the end of the project. It is also better to retain the same final document controller till the end of the project to facilitate the final document submission to the client.
- Once the final documentation procedure is approved by the client, the PCR shall prepare a master table of contents (MTOC) and have it reviewed and approved by the client.
- Based on the approved MTOC, a volume table of contents (VTOC) for each of the disciplines or volumes shall be prepared and reviewed and approved by the client.
- Once the MTOC and VTOC approvals are in place, the required stationery for the final documentation compilation must be planned and procured.
- Care should be taken to ensure that the required storage space for final documentation is also arranged to keep the documents safely.
- Vendors, such as suppliers and subcontractors involved in the piping works, must also be called, and their part of the documentation work must be discussed and agreed to by the client prior to receipt from them.
- The final document controller shall progressively compile the project's final documentation and update the final documentation monitoring status report periodically to the project management team.
- Necessary scanning or soft copy requirements of the client shall be discussed and agreed to with the client and be arranged accordingly.
- Final approval by the client on the compiled final documents is necessary and must be obtained. Any corrections, omissions and deletions related to the final documentation as commented by the client shall be attended to by the PCR quickly.
- The PCR must ensure one set of final documents, in hard or soft copy format, is obtained with the client's signature for future reference and for closing the contractual obligations, if any. For reporting of the activities, the following sample formats shall be used:

Sample Reporting Formats

Annexure – 32	:	Project Final Documentation Handover – Approval Tracking (Form No. PCR-QAF-32)

7.9 SUMMARY

As this is the final stage of the project, the project manager needs to take a proactive role to ensure that proper allocation of resources is provided at the site for various commissioning disciplines for a smoother commissioning activity. In some cases, if the commissioning scope of works is carried out by the client, then the relevant supporting engineers and technical personnel need to be provided. One of the most common mistakes in oil and gas piping projects is when the construction contractor starts to demobilize the key people from the site prior to commissioning, which will have an adverse effect on the commissioning process.

Care should be taken in addressing this issue. One more key issue at this stage of the project is the project's final documentation completion and handover to the client. Often, this aspect is underestimated and the process of final documentation completion and handover takes two to three months after the completion of the project. This should be avoided. The project manager should ensure that there is a proper final documentation team prior to start of pre-commissioning activities in the project till complete handover. Necessary resources for the final documentation team must be provided. Often, the process of final documentation is handed over to the QA/QC department as their total responsibility. This is wrong. The construction team should work hand in hand with QA/QC for successful and timely final documentation completion.

Post-commissioning

After reading this section, you should be able to understand the importance of . . .

1. Defect liability period.
2. Client notification on defects.
3. Assessment of defects and closing of defects.
4. Documentation.

© Can Stock Photo

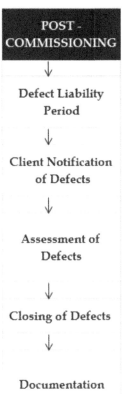

POST - COMMISSIONING

↓

Defect Liability Period

↓

Client Notification of Defects

↓

Assessment of Defects

↓

Closing of Defects

↓

Documentation

8.0 INTRODUCTION

The post-project completion phase is less critical for the contractor. However, it is very important for clients, as they need to familiarize themselves with the new project systems and attend to all the issues/problems that surface during the defect liability period (DLP). In most cases, the DLP varies from 12 months to 36 months depending on the scope of work and the criticality of the project. Many contractors end up in trouble during this period, when the costly equipment and the related systems fail. Back-to-back contract with the suppliers and subcontractors was not established, which leads to delay in attending and completing the post-project completion defects. This needs to be taken care of to achieve client satisfaction. One of the important factors is that the client's satisfaction during the DLP can lead to awarding future projects to the contractor. Hence, this needs to be given more importance. Often in the contractor's organization, there are no dedicated people to handle the post-handover of project defects during the DLP. This needs to be strengthened. Additionally, most of the clients keep 5% of the project value as a retention sum and/or performance bank guarantee to cover the defects of the DLP. This money can be easily recovered if the DLP-related defects are properly looked into. Additionally, care should be taken to ensure that all the DLP-related documents are properly maintained and handed over to the client upon successful completion of the DLP to get the certificate of completion. In this section, some of these important items are discussed in detail.

8.1 DEFECT LIABILITY PERIOD (DLP)

- As per the contract, the defect liability period (DLP) for the project as well as the scope of work need to be clearly established with the client.
- During the DLP, the defects arising in the project due to the PCR's workmanship and/or client's operational negligence and mistakes must be clearly identified.
- DLP monitoring should be carried out by the project director or contracts department and/or any person assigned by the management. They can be designated as the DLP coordinator.
- The DLP coordinator shall receive a copy of the warranty/guarantee details of the project related to pipes, valves, fittings, accessories and equipment.
- The DLP coordinator shall liaise with procurement, contracts and vendors (suppliers and subcontractors) on defects that occur during the warranty period to close the actions accordingly.
- The contracts department shall withhold the retention sum (5% or 10%) based on the criticality of the vendor's scope of work to ensure that the sum is not released until the DLP is successfully completed with the client.
- The contracts and procurement department shall ensure back-to-back warranty with the subcontractors and suppliers as against the client's DLP to ensure risk of DLP is transferred equally among the players in the project.
- Extra care should be taken to ensure proper warranty/guarantee coverage is established with vendors of critical equipment and work.
- The DLP coordinator shall maintain the records related to the defects/issues arising during the DLP for monitoring and for tackling contractual issues, if any.

8.2 CLIENT NOTIFICATION OF DEFECTS

- During the DLP, the client shall communicate the details of the defects noticed in the project through proper reports, e-mails or official letters.
- The contracts department/project director and/or DLP coordinator shall study the details of the defects notification and coordinate with the client accordingly.
- The DLP coordinator shall register the details of the defect notification from the client for monitoring and closing purposes.
- Cost/resources needed, if any, to close the defects shall also be captured in the DLP register.
- Specific to the project, a custom-generated DLP defect closing format shall be prepared and used for tracking.
- If there is provision in the client's notification to capture all the data related to closing of DLP defects, then this client format can be used.

8.3 ASSESSMENT OF DEFECTS

- For defects that need to be assessed at the site, the PCR shall send competent personnel to assess the defects. If the work is carried out by the vendor, then the respective vendor support with competent personnel shall also be sought.
- A detailed assessment report of the defects, including details, such as extent of the defect, root cause and reasons for the failure, etc., must be captured in the assessment report.
- Through discussion with the client, detailed corrective action/preventive action shall be planned and agreed to by the parties involved.
- Cost of repair and time required shall also be established in the discussion to address contractual issues, if any.
- All the members of the defect assessment team, inclusive of the client, shall sign off the assessment form, demonstrating agreement to the corrective action plan.
- All defect assessment reports shall be maintained by the DLP coordinator for tracking, monitoring and closing.

8.4 CLOSING OF DEFECTS

- The PCR shall ensure that the assessed and agreed defects are duly attended to with required quality levels to the satisfaction of the client.
- During closing of defects, it is mandatory for the client to witness and supervise to ensure proper sign-off of closing of defects.
- Details of defects rectification with photographs, data and reports (inclusive of drawings) shall be established.
- Client acceptance signature for all the defects closed shall be obtained by the coordinator to update the DLP statistics.
- Cost involved in closing the defects shall also be documented to monitor the costs against the budget.
- If the defect is due to the client, necessary charges shall be claimed from the client accordingly.

Reference Procedures

| PCR-QAP-23 | : | Defect Liability Period Procedure |

8.5 DOCUMENTATION

- The DLP coordinator shall maintain the complete history of project post-handover defects list and closing and acceptance details by the client together with cost and resources utilized.
- Photographs, drawings and pictures related to closing of defects shall be properly documented and handed over to the contracts department for closing the DLP with the client.
- The contracts department shall officially close the DLP and get the retention sum, if any is due, from the client.
- A customer satisfaction survey shall be obtained from the client to ensure satisfaction of the client and to improve the relationship further.

8.6 SUMMARY

The project director must ensure that the DLP of the project is properly handled. It is common in oil and gas construction projects that once the project is completed, the project management team is demobbed and/or the team is sent to other projects. Hence, it is very important to understand the key issues of the project during the pre-commissioning and commissioning stages and address some of the issues that might crop up during the post-commissioning period of the project. Proper maintenance of project-related documentation till the end of the DLP is very important, as the client might ask for vital information on critical areas related to material and/or equipment. Failure to keep such information handy may lead to additional expenses for the contractor. In many cases, during the DLP, additional work and/or rework due to client operations may pop up and a proper solution provided by the contractor will yield revenue as well as confidence to the client to award future projects. DLP-related documentation needs to be maintained to ascertain knowledge and lessons learned in the project.

Project management

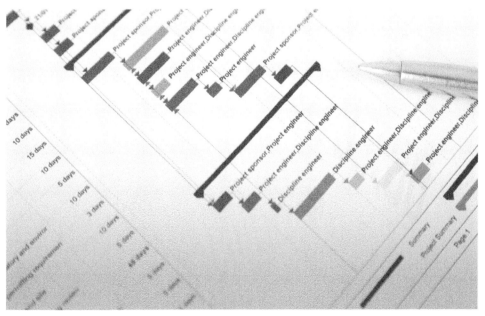

After reading this section, you should be able to understand the importance of . . .

1. Project management in oil and gas projects.
2. Factors impacting project performance of oil and gas projects.
3. Relative importance of project- and organizational-related factors on oil and gas project performance.

© Can Stock Photo

9.0 INTRODUCTION

Every project is different. Success in each and every project undertaken is becoming an impossible thing in today's context. Specifically, in the oil and gas construction industry, there are stories of many successful companies as well as failures. Project management is getting more complicated every passing day. Success in one project does not guarantee the same type of success in another similar project. Many variables and factors play an important role in determining the success of projects. Most contractors have an idea as to what the important factors that can contribute significantly to project success are and are not. It is important to understand this because it will help to focus on that specific area. Specifically, in this section, factors that are important for oil and gas projects and that significantly impact project performance for oil and gas projects in Malaysia are discussed in detail. This will definitely help the contractors to understand, focus and move forward. While both project- and organizational-related factors mentioned in this section are important for the oil and gas industry, the most important project-related factors and organizational-related factors are identified and explained.

Success of a project in the oil and gas industry depends on many factors, which impact performance from diverse angles. Some factors are unique to the project itself, while others are unique to the organization. Similarly, some factors are unique to the oil and gas industry itself, while others are unique to the country in which the project is undertaken and/or the organization in which it operates.

Recent research studies on the oil and gas construction industry's project performance in Malaysia have revealed some interesting factors, which are unique to the country in particular, and to the oil and gas industry in general. The research has focused on two distinctive sets of factors that are likely to impact project performance, i.e., the "hard factors, commonly known as project-related factors" and the "soft factors, commonly known as the organizational factors."

Key project performance dimensions of oil and gas projects, such as "time, cost, quality, safety and financial" dimensions, were studied in this research.

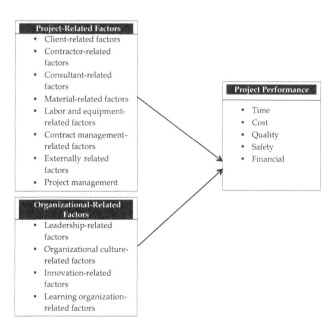

The definitions for each of the factors mentioned above in the research framework and the keywords are as follows:

Project – A project is a temporary endeavor undertaken to create a unique product or service. "Temporary" means every project has a definite beginning and definite end. "Unique" means the product or service is different in some distinguishing way from all other products or services. Projects are often implemented as a means to achieve the organization's strategic plans.

Project management – Can be defined as the application of a collection of tools and techniques to plan, control and direct the use of diverse resources for the accomplishment of a unique, complex, one-time task within the time, cost and quality constraints. Each task requires a particular application of tools and techniques structured to fit the task environment and life cycle of the task (from concept to completion).

Client-related factors – Factors such as finance and payment of completed work in the project, too much owner interference on project matters, slow decision making on critical issues of the project and unrealistic duration imposed for completion by the owner.

Contractor-related factors – Factors related to subcontractors of the project, such as improper site management, improper planning and work execution, inadequate experience, mistakes during construction, improper construction methods and delays caused by the subcontractors.

Consultant-related factors – Factors such as technical items related to contract management, preparation and approval of project drawings, quality assurance/control of work carried out in the project, long waiting time for approval of tests and inspections by the consultant team.

Material-related factors – Factors, such as delay in delivery of materials and equipment, wrong delivery of materials and equipment, short supply of materials and equipment, quality and performance issues of supplied material and equipment.

Labor and equipment-related factors – Factors, such as labor supply, labor shortage, labor productivity, labor skill issues, equipment availability and equipment failure issues faced during the project.

Contract management-related factors – Factors, such as change orders imposed by the client, mistakes, discrepancies and technical inconsistencies in contract documents, disputes and negotiations during construction related to time, cost, quality and technical matters.

Externally related factors – Factors such as weather conditions, changes in regulations and social, political, religious and economic changes that happen during the course of the project.

Project management tools/techniques-related factors – Project management is challenging, with many complex tasks, objectives and responsibilities. There are many tools/techniques available to assist in accomplishing the tasks and executing the responsibilities to meet the objectives. Some require a computer with supporting software, while others can be undertaken manually. Project managers should choose a project management tool/technique that best suits the project undertaken. No one tool/technique will address all project management needs. The program evaluation review technique (PERT) and Gantt charts are two of the most commonly used project management tools.

Time overrun – Delay in completing the project within the agreed time duration of the project due to factors such as inadequate planning by contractors, improper site management by the contractors, inadequate project handling experience of contractors and delay in payments by the client for the work completed by the contractors.

Cost overrun – Factors related to the contract, such as change orders (changes to the original deliverables and requirements, mistakes and discrepancies in the contract document) that result in cost of execution of the project exceeding the estimated cost of the project.

Leadership – Leadership is different from management. Leadership and management are two distinctive and complementary systems of action. Each has its own functions and distinctive features. Management is about coping with complexity. Leadership, by contrast, is about coping with change. Leaders do not make plans, they do not solve problems and they do not even organize people. What they do is prepare the organization for change and help the people cope as they struggle through it.

Organizational culture – Organizational culture can be defined as the set of values, beliefs and behavior patterns that form the core identity of an organization, which help shape the employees' behavior. It provides the selection mechanisms or norms and values that people enact and perform.

Innovation – Innovation is defined as the firm's intellectual capability to produce new products to the market to sustain and to improve organizational performance. Innovative organizations are successful and lead the market in which they operate. R&D and adopting new technologies are the keys to innovate new products or services in organizations.

Learning organization – A learning organization is an organization skilled at creating, acquiring and transferring knowledge and at modifying its behavior to reflect new knowledge and insights in all its business processes for sustainability and for improvement in its performance.

A total of 82 oil and gas experienced professionals from various oil and gas construction companies in Malaysia participated in this research. The collected data were scientifically tested for reliability and were found reliable for further analysis. The data were analyzed to find out the relative importance of these project-related factors and organizational-related factors on oil and gas project performance.

The results are as follows:

9.1 RELATIVE IMPORTANCE OF FACTORS IN OIL AND GAS PROJECTS

Factors	Relative Importance Ranking
Project-related factors	
Client-related factors	7
Contractor-related factors	3
Consultant-related factors	4
Material-related factors	2
Labor and equipment-related factors	1
Contract management-related factors	6
Externally related factors	8
Project management tools/techniques-related factors	5
Organizational-related factors	
Leadership-related factors	3
Organizational culture-related factors	2
Innovation-related factors	4
Learning organization-related factors	1
Project Performance	
Time-related performance	4
Cost-related performance	5
Quality-related performance	3
Safety-related performance	1
Financial-related performance	2

N = 82

From the above analysis, it is evident that the most important factors for project performance of the oil and gas construction sector are as follows:

Oil and gas construction sector's project performance – most important factors

Variable	Most important factors
Project-related factors	Labor and equipment, material and contractor-related factors
Organizational-related factors	Learning organization, organizational culture and leadership-related factors
Project performance dimensions	Safety, financial and quality-related dimensions

The above analysis indicates that for oil and gas projects, the most predominant project-related factors over the other factors are labor and equipment, material and contractor-related factors. These three hard factors are really vital for oil and gas industry projects because of the following reasons:

1.	Labor and equipment	–	Oil and gas labor and equipment are specialized in nature compared to conventional projects. The cost of oil and gas labor and equipment is high compared to conventional construction projects.
2.	Material	–	Due to the corrosive nature of the seawater and the chemicals used in the oil and gas pipeline materials, the materials used in oil and gas projects are special in nature and expensive. Hence, the material-related factors are really important.
3.	Contractor	–	Due to the oil and gas projects' location, quality of work and safety requirements needed in the project are very stringent; the contractors who carry out the work are very important for the projects' success. As everyone is aware, there are many differences between oil and gas contractors and conventional project contractors.

Similarly, with respect to the organizational-related factors, the most important factors that emerged in this study can be explained as follows:

1.	Learning organization	–	As the cost of carrying out oil and gas projects is naturally high due to location and specialty requirements, any mistake in the project will lead to cost escalation. In order to prevent the recurrence of such mistakes, it is most common that in the oil and gas industry the lessons learned become part and parcel of the project management's learning techniques. This will help to prevent costly mistakes and motivate a learning organization culture in the organization as a whole.
2.	Organizational culture	–	Compared to the conventional construction industry, oil and gas construction organizations as well as their projects must nurture a different culture, where safety, quality and planning are given utmost priority. The culture and values practiced by the oil and gas construction project/organization must be entirely different, as the cost of mistakes and their consequences are really severe. Compliance and conformation are of topmost priority in all the activities.

3.	Leadership	– The leadership qualities required in an oil and gas project are really critical, as the drive and determination to get things done begin with the leader of the project. As the oil and gas project execution needs different skill sets with a high level of technological interface, a mature leadership quality is essential to achieve the desired success in the project.

In the case of project performance dimensions for oil and gas industry projects, the research study concludes that safety, quality and financial performance are relatively important over time and cost performance. Traditionally, even for oil and gas projects, time and cost performance are considered as very important. However, this recent study clearly indicates that safety, quality and financial performance are relatively important over time and cost performance. There must be a paradigm shift in the eyes of the oil and gas project performance practitioners.

9.2 SUMMARY

The project director and project manager of oil and gas piping projects should understand the important factors that have an impact on project performance. More attention should be paid to the most important factors, such as labor and equipment, material, contractor, learning organization, leadership and organization culture, compared to the other areas, to achieve desired project performance. As the above factors have been identified in this research, the reliability of the results is good and supports project management. Where necessary, the related information can be disseminated among the team members to create awareness towards achieving better performance. As every project is unique and the team is different, it is important to ensure that key factors that affect project performance are well understood. KPIs, objectives and controls must be established in these areas to prevent unexpected outcomes.

ISO standards

After reading this section, you should be able to understand the importance of . . .

1. ISO standards in oil and gas project management.
2. ISO 9001 quality management systems.
3. ISO/TS 29001 oil and gas sector-specific quality management systems.
4. ISO 14001 environmental management systems.
5. OHSAS 18001 occupational health and safety management systems.
6. ISO 21500 project management systems.

© Can Stock Photo

10.0 INTRODUCTION

Success and failure of any organization and its projects largely depends on their systems and practices. Over the years, many management concepts have been introduced and practiced around the world. However, application of most of the management system tools is very difficult for construction companies globally due to the temporary nature of the project. However, in the recent past, it has been found that the application of ISO 9001, ISO 29001, ISO 14001, OHSAS 18001 and ISO 21500 systems in construction industry projects has led to improved results. Because of this, the statutory bodies in the ISO member countries have made it mandatory to implement some of these ISO systems in the organizations to achieve better results. Many of the critical control points related to quality, environment, occupational health and safety, oil and gas sector-specific quality requirements and overall project management are explained in these standards. Proper implementation of these standards will help the organization to achieve reasonably good benefits. In this section, the abovementioned five important standards are explained in brief.

For effective piping discipline-related scope of work completion in oil and gas projects, the following ISO standards can be implemented and followed. Proper implementation of these ISO standards will help the organization to overcome all the project-related issues, such as time overruns, cost overruns, quality issues, safety issues, client complaints, material wastage, rework, etc.

a) ISO 9001 quality management system.
b) ISO/TS 29001 petroleum, petrochemical and natural gas industries – sector-specific quality management systems – requirements for product and service supply organizations.
c) ISO 21500 project management systems.
d) ISO 14001 environmental management systems.
e) OHSAS 18001 occupational health and safety management systems.

10.1 ISO 9001 QUALITY MANAGEMENT SYSTEM

This is one of the most popular international standards in the ISO family of standards. Implementing this standard in the project will help to ensure that all the processes of the project as well as the parent organization are properly governed by documented procedures and protocols. Enhancement of product/service quality is very much focused on in this standard, with the aim of achieving client satisfaction. Responsibility, authority and accountability are properly documented with quality objectives (KPIs) for each process. A complete transformation of the organization/project can be expected by implementing this standard with top management's commitment. Some of the key aspects of this standard are as follows:

a) Focusing on client satisfaction
b) Identification of key processes
c) Identification of risks and mitigation measures for each key process
d) Leadership and commitment
e) Interested parties' identification and management
f) Internal and external issues' identification and mitigation
g) Quality policy
h) Quality objectives (KPIs)
i) Quality manual
j) Quality procedures for all the key processes
k) Documented information
l) Internal audit and management review
m) Continuous improvement
n) Resources provision

10.2 ISO/TS 29001 OIL AND GAS SECTOR-SPECIFIC QUALITY MANAGEMENT SYSTEM

This international standard is specifically meant for oil- and gas-related organizations, projects and service providers. This international standard is an advanced version of the ISO 9001 standard with additional features/requirements imposed by taking into account the requirements of the oil and gas projects. Implementing this standard will help the project to ensure that all the client's requirements are properly met. Some of the key requirements of this standard, in addition to the ISO 9001 standard, are listed below:

1. All the project-related and/or business-related documents must be maintained for a minimum period of five years to ensure identification/traceability.
2. Once the project is completed, post-delivery services (defect liability period), warranty and guarantee-related items become the responsibility of the product/service provider to the client.

3. Once the product/service is delivered and/or the project is completed, post-delivery performance of the project/product/service must be monitored.
4. If project-related works are subcontracted by the main contractor, then the responsibility and accountability of the subcontracted works shall be that of the main contractor.

10.3 ISO 21500 PROJECT MANAGEMENT SYSTEM

ISO 21500:2012
Guidance on Project Management

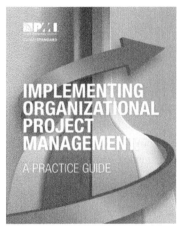

This international standard is an excellent guide to project management. Irrespective of the nature of the project, it holistically covers all aspects of project management. The four key phases of project management, i.e., initiating, planning, controlling and closing of the project, are embedded in the project management fundamentals and protocols, which are essential to ensure success. The complete standards based on the key inputs from the Project Management Body of Knowledge (PMBOK) are available in a handbook published by Project Management Institute (PMI). Some of the key inputs for better project governance advocated by this standard for successful project management are as follows:

a) Requirement of a project charter
b) Requirement of a project execution plan
c) Requirement of a construction management plan
d) Requirement of a subcontracting management plan
e) Requirement of a communication plan
f) Requirement of a project risk identification and mitigation register
g) Requirement of a stakeholders' management plan
h) Requirement of a procurement plan
i) Requirement of a project schedule
j) Requirement of a project budget and cost control plan

In most cases, some or many of the above plans are not available at the time of project execution and that is the main reason for many of the issues in projects. Proper implementation of this international standard with the top management's commitment will lead to project success.

10.4 ISO 14001 ENVIRONMENTAL MANAGEMENT SYSTEM

Projects carried out currently and in upcoming years need to pay more attention to proper environmental management. This international standard emphasizes the need for companies and projects to implement necessary environmental management systems to protect the environment and to ensure reduced land, water, air and noise pollution. Some of the salient features of this standard require the projects to establish the following documents relevant to the project/organization, which will help in achieving a good environmental management system for its operations.

a)   Environmental policy
b)   Environmental objectives (KPIs)
c)   Environmental management plan
d)   Environmental procedures
e)   Operational control
f)   Aspects and impacts identification/mitigation
g)   Compliance to legal and statutory requirements
h)   Top management commitment
i)   Continuous improvement

10.5 OHSAS 18001 – OCCUPATIONAL HEALTH AND SAFETY MANAGEMENT SYSTEM

This international standard emphasizes the need for projects/organizations to have a proper occupational health and safety management system in their operations. Specifically, for projects, such as oil and gas projects, where safety is considered as the topmost priority, implementation of this standard will help the project to prevent safety and health-related problems. Now and in the days to come, more attention should be paid to safety and health issues at the workplace. Hence, adhering to this standard will help the project/organization to fully meet the client's requirements. Some of the requirements of this standard are:

a)   Occupational health and safety (OHS) policy
b)   OHS objectives (KPIs)
c)   OHS manual
d)   OHS procedures
e)   Hazard identification and mitigation plan
f)   Operational controls
g)   Accident/incident investigation, correction, corrective action
h)   OHS training
i)   OHS audits and management reviews
j)   Continual improvement

10.6 PROCESS OF GETTING ISO CERTIFICATION

Construction organizations and projects that would like to achieve the relevant ISO certification discussed above should embark on this initiative with the assistance of their internal QA/QC manager and/or external consultants. A number of training sessions on the respective ISO standard awareness, documentation and internal audit must be conducted, either for a group of people or for selected people of the organization or project. Required documentation for the respective ISO standard as stipulated in the standard must be prepared and practised. Evidence of ISO standard implementation and effectiveness must be demonstrated. Once satisfactory preparatory works are in place in the project or organization, then the ISO certification bodies, such as Bureau Veritas, TUV, DNV, Lloyd's Register, etc., can be contacted to get the ISO certification done for the organization/project.

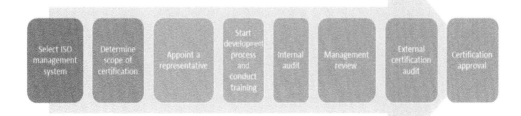

10.7 SUMMARY

Construction companies that are involved in oil and gas piping and related construction projects should implement the ISO 9001, ISO/TS 29001, ISO 14001, OHSAS 18001 and ISO 21500 management systems. Almost 90% of the management-related requirements in oil and gas projects are covered in these five standards; hence, compliance to these standards gives merit to the organizations right from tendering to getting contracts. These standards will also help the construction companies to ensure trouble-free implementation of quality and oil- and gas-specific requirements as well as ensure environmental, safety and acceptable project management practices in the projects. Top management personnel in construction organizations need to ensure that these standards are implemented in the projects success-fully. Implementation of these standards can help the company to grow as well as identify all the potential problems in the organization to ensure timely corrective and preventive actions. Regulatory officials and clients will value the organization to a greater extent if the organization is certified to these standards. One important caution is that many construction companies, after getting certified to these standards, do not pay attention to practical imple-mentation at project sites, which often leads to issues recurring in the project, in turn leading to time overruns, cost overruns, quality issues, safety issues and customer complaints. These should be avoided at all costs.

Appendices

PIPING CONTRACTOR'S COMPANY LOGO	THIRD-PARTY INSPECTION REQUEST FORM (RFI) [Insert project title here]	CLIENT LOGO

To	:		RFI No.	:	
Requested By	:		Date	:	

Please be invited to attend the following inspection according to the following schedule:

Planned inspection date/time	:	
Inspection location	:	
Inspection subject	:	

INSPECTION DETAIL

Discipline Involved	☐ Mechanical/Piping ☐ Civil & Structural ☐ Electrical	☐ Instrument ☐ Process ☐ Pipeline	☐ Others - Specify
Inspection/ Purpose	☐ Witness	☐ Joint Inspection	☐ Close Out NCR
Reference	☐ Drawing ☐ Specifications ☐ Others	Drawing No. : Specs No.:	

The above document(s) is(are) attached: ☐ Yes ☐ No

Requested by: [PCR Representative]		Acknowledged by: [Client Representative]	
Name	:	Name	:
Designation	:	Designation	:
Date	:	Date	:
Signature	:	Signature	:

INSPECTION STATUS (To be completed after the inspection by PCR or Client)

Actual Inspection Date/Time :

☐ The testing is satisfactory and works may proceed to the next stage

☐ Checklist Ref. No.:

☐ The above activities need to be re-inspected on :

☐ Another RFI has been issued for the above activity.   New RFI No.:

☐ Not Approved due to :

REMARKS

Requested by: [PCR Representative]		Acknowledged by: [Client Representative]	
Name	:	Name	:
Designation	:	Designation	:
Date	:	Date	:
Signature	:	Signature	:

Note:
- *RFI is valid for 48 hours (working day) from the RFI issue date. If the inspection is not conducted within 48 hours, a new RFI is to be issued.*
- *Please make sure that RFI is linked to Inspection Report and NCR.*

PIPING CONTRACTOR'S COMPANY LOGO	SOURCE INSPECTION REPORT	CLIENT LOGO

Project	:		Date of Inspection	:	
Location	:				
Inspected By	:		Report No.	:	

Category

☐ Incoming   ☑ In Process   ☐ Final Inspection   ☐ Site Inspection

INSPECTION DESCRIPTION : FINAL VISUAL & DIMENSIONAL FOR ITEM...

Please write details of inspection and/or attach photos of the inspected items.

INSPECTION STATUS

Item Description	:
Quantity Inspected	:
Total Accepted	:
Code, Standards and Drawing	:
Remark	:

Inspected by:	Reviewed by:
Name :	Name :
Designation :	Designation :
Date :	Date :
Signature :	Signature :

PIPING CONTRACTOR'S COMPANY LOGO	MATERIAL RECEIPT INSPECTION & TEST REPORT

PROJECT: _____ REPORT NO. : _____
_____ P.O. NO. : _____
_____ D.O. NO. : _____

ITEM	MATERIAL DESCRIPTION	QTY	UNIT	REMARKS

MANUFACTURER: SUPPLIER: UNLOADED : _____ HRS
DATE :

EX. VESSEL / AIRCRAFT   TRUCK / TRAILER NO.   DOCS. OK YES ☐ NO ☐   DEFECT REPORT NO.   STORAGE AREA

MATERIAL DEFECTS : YES ☐ (MATERIAL DAMAGED / MISSING / WRONG/...) NO ☐

MANUFACTURER'S TEST CERTS. PROVIDED : YES ☐ NO ☐ COMPLIANCE : YES ☐ NO ☐

VISUAL / DIMENSIONAL / TEST DETAILS

REMARKS:

ACTION	PIC	NAME	SIGNATURE	DATE
PREPARED BY	STORE KEEPER			
INSPECTED BY	ENGINEER			
REVIEWED BY	QA/QC MANAGER			
VERIFIED BY	CLIENT			

PIPING CONTRACTOR'S COMPANY LOGO		LIST OF CRITICAL MATERIAL REQUIRED IDENTIFICATION & TRACEABILITY			
ITEM NO.	DESCRIPTION OF PRODUCT	METHOD OF IDENTIFICATION & TRACEABILITY	TRACEABILITY DATA	EXTENT OF TRACEABILITY	RECORD FORMAT

PIPING CONTRACTOR'S COMPANY LOGO	LIST OF AUTHORIZED PERSONNEL FOR WITHDRAWING MATERIAL FROM THE STORE

PROJECT TITLE: _____

NO.	NAME	SIGNATURE SPECIMEN	MATERIAL ALLOWED TO BE WITHDRAWN

	PREPARED	APPROVED
Signature		
Name		
Designation		
Date		

PIPING CONTRACTOR'S
COMPANY LOGO

STOCK CARD

Item:			Ref. No.:					Stock Card No.:
DATE	Reference	IN	OUT	Balance	Remarks	DATE	Reference	IN	OUT	Balance	Remarks

Item:			Ref. No.:					Stock Card No.:

STORE ISSUE NOTE

PIPING CONTRACTOR'S COMPANY NAME
Address:
Phone Numbers:

NO.:

PROJECT : _____

REQUESTED BY : _____

LOCATION : _____ DATE : _____

Item	Description	Unit	Qty	Remarks

DATE : _____ RECEIVED BY : _____

ISSUED BY : _____ NAME : _____

PCR-QAF-07

PIPING CONTRACTOR'S COMPANY LOGO	NON-CONFORMANCE REPORT (NCR)

NCR NO.: _____

Please tick on appropriate box

A) NC CATEGORY: ☐ PROCESS   ☐ PRODUCT

B) OCCURRENCE: ☐ 1ST OCCURRENCE   ☐ RECURRENCE (REPEATED)

C) NC FOUND / DISCOVERED AT:

☐ INCOMING   ☐ IN PROCESS   ☐ FINAL   ☐ SUPPLIER / SUB-CON   ☐ OTHERS _____

D) REQUIREMENT (Indicate one section or clause # from relevant Standard, Procedure, Work Instruction, Drawing, etc.)

E) DETAILS OF NON-CONFORMANCE *(state N/A if not applicable)*

PROJECT _____   CLIENT   : _____

NCR To : _____   DEPARTMENT: _____

DESCRIPTION OF NONCONFORMANCE:

	Issue	Approve

F) DISPOSITION OF NONCONFORMING PRODUCT *(to be filled up by NCR owner)*

		Incharge	Review	Approve
☐	Use As Is - *Justification*:			
☐	Screen & Use			
☐	Rework / Repair (issue RFI)			
☐	Reject/Dispose			
☐	Return to Vendor			
☐	Other - *Please specify*:			

G) INVESTIGATION AND ACTION *(to be filled up by NCR owner)*

1. ROOT CAUSE:

2. CORRECTION:

3. CORRECTIVE ACTION:

H) VERIFICATION BY QA/QC DEPARTMENT

1) DISPOSITION RESULT *(to be filled up by Issuer)*

☐ ACCEPTED   ☐ NOT ACCEPTED

2) CORRECTIVE ACTION TAKEN *(to be filled up by Issuer)*

☐ SATISFIED   ☐ NOT SATISFIED

3) CONCLUSION *(to be filled up by QA/QC Manager/Lead or verifying person)*

☐ CLOSED   ☐ RE-ISSUE NEW NCR

COMMENT:

	Issue	Approve

PIPING CONTRACTOR'S COMPANY LOGO	DOCUMENT TRANSMITTAL

	Transmittal No.:

Send To: Transmittal Date:

                                                      Method:

Subject:

Attention:

Items Transmitted (Tick where applicable)					
Catalogues		Inspection/Test Reports		Reports	
Certificates/Calibration		Meeting Minutes		Schedules	
Calculations		Method Statement		Sketches	
Drawings		Procurement		Specifications	
Fax/Letters		QA Documents		Tender Documents	
				Others	

NO.	REV. NO.	DOCUMENT NO.	TITLE/DESCRIPTION	MEDIUM	QTY

Medium Key:

Medium A- Hard Copy   Medium B- PDF Drawing   Medium C- Soft Copy

Status:

	For Action		For Construction		For Review		For Information
	For Comment		For Approval		For Reference		Others

Return To:

Document Control Center

Piping Contractor Company Name

Address:

Phone No:

Fax:

Issued by

Name:

Date:

ACKNOWLEDGEMENT	
ISSUANCE	RETRIEVAL
We acknowledge receipt of the above transmittal	We acknowledge retrieval of the drawing(s)/document(s)
Signature:	Signature :
Name :	Name :
Date :	Date :

FIT-UP AND VISUAL INSPECTION REPORT (PIPING)

PIPING CONTRACTOR'S COMPANY LOGO

Title:		Contract No.:
		PCR Job No.:
		Date :

Client:
Contractor:
Report No.:
Specification :

Sl. No.	Drawing No.	Rev. No.	Line No./ Spool No.	Joint No.	Weld Type	Size/ Schedule/ Thickness (mm)	Pipe/Heat No. *	WPS No.	Welder's ID				Visual Inspection		NDT Required							Remarks
									Root	Fill	Fill	Cap	Fit-up (Accept/ Reject)	Final (Accept/ Reject)	RT	UT	MT	PT	PMI	PWHT	Hardness	

* Legend: P - Pipe   F - Flange   E - Elbow   T - Tee   R - Reducer   B - Branch Fitting

Consumable Batch Detail

Type	Brand	AWS Class	Size	Batch No.

	Contractor	Third Party	Client
Signature			
Name			
Position			
Date			

PCR-QAF-10

PIPING CONTRACTOR'S COMPANY LOGO	DIMENSIONAL REPORT	

Client:	Title:	Contract No:
Contractor:		PCR Job No.:
Report No.:	Specification:	Date:
Reference Drawing:		Item Location/Area:

MEASURING POINT	DRAWING REQUIRE-MENT (mm)	TOLERANCE	ACTUAL MEASUREMENT (mm)	METHOD/EQUIPMENT	DATE	REMARKS

RESULT	CONTRACTOR	CONTRACTOR	CLIENT
☐ Accept ☐ Reject	Signature: Name: Date:	Signature: Name: Date:	Signature: Name: Date:

WELD SUMMARY SHEET (PIPING)

PIPING CONTRACTOR'S COMPANY LOGO

Client :
Contractor :
Report No. :
Drawing No. :

Title :
Specification :

Contract No. :
PCR Job No. :
Date :
Test Pack No. :

LINE NO./SPOOL NO.	JOINT NO.	SIZE /SCH	WPS NO.	WELD PROCESS	WELDER NO.		FIT-UP & VISUAL REPORT NO.	NON-DESTRUCTIVE TEST									PMI REPORT NO.	FERRITE CONTENT REPORT NO.	PWHT REPORT NO.	RT AFTER PWHT	PAINTING REPORT NO.	REMARKS
					ROOT	CAP		RT		UT		MPI		DPI							
								Report No.	Result	Report No.	Result	Report No.	Result	Report No.	Result						

	Contractor	Contractor	Client
Signature			
Name			
Position	QC Inspector	QA/QC Engineer	
Date			

PCR-QAF-12

Page 1 of 1

PIPING CONTRACTOR'S COMPANY LOGO	RELEASE NOTE FOR PAINTING

Client:	Title:	Contract No.:
Contractor:		PCR Job No.:
Report No.:	Specification:	Date:

SI No.	Drawing No.	Spool No.	Item/Description	Quantity	Remarks

It is confirmed that all the above items are:

☐ Visually inspected and accepted
☐ NDE carried out as applicable and accepted

	Contractor	Contractor	Contractor
Signature			
Name			
Position	Fabrication Supervisor/Foreman	QC Inspector Welding	QC Inspector Painting
Date			

PIPING CONTRACTOR'S COMPANY LOGO	BLASTING AND PAINTING INSPECTION REPORT

Client:	Title:	Contract No.:
Contractor:		PCR Job No.:
Report No.:	Specification:	Date:

Climatic Condition

	Blasting	Primer	2nd Coat	3rd Coat			Remarks/Comments
Date							
Time (Hours)							
Dry Point ($^{\circ}$C)							
Wet Point ($^{\circ}$C)							
Relative Humidity (%)							
Dew Point ($^{\circ}$C)							
Steel Temp. ($^{\circ}$C)							
Weather							

Surface Preparation / Painting Details

Method		Method	Spray / Roller / Brush
Surface Profile		Painter	
Standard		Others	
Type Abrasive			

Coating System

No. of Coats	Name of Product	Paint Generic Name	Color	DFT
			Total	

Product Name	Batch No. (Part A)	Batch No. (Part B)	Batch No. (Thinner)

Average Dry Film Thickness (Micron)

Item/Description	Primer	2nd Coat	3rd Coat

Instruments Used

Item/Description	Serial No.	Date of Calibration	Others

Remarks:

	CONTRACTOR	PAINT MANUFACTURER	CLIENT
Signature			
Name			
Position	QC Inspector - Painting		
Date			

| PIPING CONTRACTOR'S COMPANY LOGO | PAINTING/COATING FINAL INSPECTION REPORT |

Client:	Title:		Report No.:
Contractor:	Specification:		Date:
Job/Item Description:			
Reference Drawings:			
Location:			
Painted Surface:	Internal/External		
Client Paint Specification:			
Surface Preparation:			
Painting/Coating Type:			
No. of Coats Applied:			
Visual Inspection:			
Top Coat DFT			
Degree of Cure:			
Pinhole Testing:			
(Wet Sponge Method)			
Holiday Detection:			
(Brush/Coil)			
K.V Setting (for Pipe Line)			
Remarks:			

	Contractor	Paint Manufacturer	Client
Signature			
Name			
Position	QC Inspector - Painting		
Date			

INTERNAL PACKING LIST

PIPING CONTRACTOR'S COMPANY LOGO

Project Name:	

Name:		SHIPMENT NO.:	
Designation:		Other Documents:	
Tel. No.:	Date:		

Ship To:	
Address:	

Person In Charge:

Postal Code:	State:
Phone No.:	Attn. To:
Fax No.:	Email:

Phone No.: Email:

Load out Date:

Type of Transport:
- [ ] Short Distance Truck
- [ ] Long Distance Truck
- [ ] Sea Transport
- [ ] Air Transport (as required)

Vehicle No./IMO: _____

Type of Storage:
- [ ] Outdoor
- [ ] Indoor

Special Requirements: _____

NO.	DESCRIPTION	QUANTITY

Total Number of Parts/Sub-unit:		Remarks:
Number of Layer:		
Total Number of Parts:		
Outer Dimensions L x W x H in mm:		
Total Weight in kg:		
Stacking Factor:		

PIPING CONTRACTOR'S LOGO

INTERNAL PACKING LIST

Picture of Parts/Materials:

	Prepared	Reviewed	Approved
Name: Designation: Date: Signature:	Project Engineer	Project Manager	Director

	QA/QC Representative	Client's Representative
Name: Designation: Date: Signature:		

WELDING PROCEDURE SPECIFICATION

WPS NO. : _____ REV. NO. : _____ REV.DATE. : _____ CODE : _____ SHT. ___ OF ___

PROCESS : _____ TYPE : MANUAL _____ SEMI-AUTO _____ MACHINE _____ AUTO ___

WELD GROOVE : SINGLE VEE _____ DOUBLE VEE _____ SINGLE BEVEL _____ DOUBLE BEVEL ___

SINGLE U _____ DOUBLE U _____ SINGLE J _____ DOUBLE J ___

COMB. ANGLE GROOVE _____ GROOVE ANGLE <30° _____ T,K & Y _____ SQUARE BUTT ___

FILLET WELDS : SINGLE PASS _____ MULTI-PASS _____

SUPPORTING PQR NO : (1) _____ (2) _____ (3) _____

MATERIAL GROUPING : ASME: P. NO : _____ GROUP NO. _____ TO P. NO : _____ GROUP NO. _____

AWS GROUP NO. : _____ TO GROUP NO. : _____

GRADE : _____

TUNGSTEN ELECTRODE : SIZE : _____ TYPE : _____

SHIELDING GAS TYPE : YES _____ NO _____ GAS _____ MIX % _____ FLOW _____ LPM

PURGING GAS TYPE : YES _____ NO _____ GAS _____ MIX % _____ FLOW _____ LPM

TRAILING GAS TYPE : YES _____ NO _____ GAS _____ MIX % _____ FLOW _____ LPM

FILLER METAL SPECIFICATION :

( ) _____ MFG / BRAND _____

( ) _____ MFG / BRAND _____

FILLER METAL CLASSIFICATION :

( ) _____ A NO. _____ F NO. _____

( ) _____ A NO. _____ F NO. _____

WIRE / FLUX CLASSIFICATION : _____ FLUX BRAND : _____

WELDING CURRENT DC _____ AC _____ POSITIVE _____ NEGATIVE _____ PULSED _____

& POLARITY : DC _____ AC _____ POSITIVE _____ NEGATIVE _____ PULSED _____

WIRE FEED SPEED RANGE, cm/min. : (FCAW-GS) _____ (SAW) _____ (GMAW) _____

MODE OF METAL TRANSFER FOR GMAW : SHORT CIRCUIT : _____ GLOBULAR : _____ SPRAY : _____

SINGLE OR MULTIPLE ( ) SGL. _____ M'TPL. _____ STRING _____ WEAVE _____ SPLIT LAYER _____

PASS USING STRING ( ) SGL. _____ M'TPL. _____ STRING _____ WEAVE _____ SPLIT LAYER _____

WEAVE OR SPLIT LAYER : ( ) SGL. _____ M'TPL. _____ STRING _____ WEAVE _____ SPLIT LAYER _____

WELD WIDTH MAXIMUM : MANUAL _____ MM SEMI-AUTO _____ MM MACHINE _____ MM

VERTICAL PROGRESSION : ( ) UP _____ DOWN _____

( ) UP _____ DOWN _____

POSITION OF PRODUCTION WELD : SMAW _____ GMAW _____ FCAW-GS _____ GTAW _____ SAW _____

NO. OF ARCS : _____ ELEC. SPACING: LONGITUDINAL _____ LATERAL _____ ANGLE _____ TRAVEL DIRECTION _____

ROOT WELD BACKING : NO _____ YES _____ BACKING MATERIAL _____

STICK OUT LENGTH : _____ MM ORIFICE / CUP SIZE : _____ CONTACT TUBE TO WORK DISTANCE : _____ MM

MAX. NO. OF HEAT CYCLES : _____ MAX. TIME BETWEEN HEAT CYCLE : _____

TREATMENT OF BACKSIDE OF ROOT WELD : NONE _____ GRIND _____ GOUGE _____ BRUSH _____ OTHERS _____

PREHEAT / INTERPASS TEMPERATURE : MINIMUM _____ MAXIMUM _____

METHOD OF PREHEATING : _____

METHOD OF TEMPERATURE CHECK : _____

POST WELD HEAT TREAT REQUIRED : NO _____ YES _____ TEMP.: _____ HOLDING TIME : _____

CHARPY IMPACT TESTING REQUIRED : NO _____ YES _____ REPORT NO. _____

HARDNESS TESTING REQUIRED : NO _____ YES _____ REPORT NO. _____

TENSILE TEST : NO _____ YES _____ REPORT NO. _____

BEND TEST : NO _____ YES _____ REPORT NO. _____

MACRO-ETCH & PHOTO : NO _____ YES _____ REPORT NO. _____

NDT : NO _____ YES _____ REPORT NO. _____

CHEMICAL COMPOSITION : NO _____ YES _____ REPORT NO. _____

RANGE OF WELD METAL THICKNESS :

( ) MIN. _____ MAX. _____

( ) MIN. _____ MAX. _____

( ) MIN. _____ MAX. _____

RANGE OF PIPE/TUBULAR DIAMETERS :

MIN. : _____ MAX. : _____

RANGE OF BASE METAL THICKNESS WHEN PRODUCTION WELD NOTCH TOUGHNESS IS:

	REQUIRED	NOT REQUIRED
MIN.	_____	_____
MAX.	_____	_____

WELDING PROCEDURE SPECIFICATION												
WPS NO. :				REV. NO. :		REV. DATE :				SHT.2 OF 2		
SPECIFIED PRODUCTION WELDING PARAMETERS												
PASS/ WELD LAYER	WELDING PROCESS	ELECTRODE/FILLER METAL CLASSIFICATION	ELECTRODE/FILLER METAL DIAMETER (mm)	CURRENT TYPE AND POLARITY	AMPERAGE RANGE (A)	VOLTAGE RANGE (V)	TRAVEL SPEED RANGE (mm/min.)	HEAT INPUT RANGE (kJ/mm)	STRING	WEAVE	SPLIT LAYER	MAXIMUM INTERPASS TEMPERATURE (°C)

JOINT PREPARATION

CONTRACTOR :	THIRD PARTY / CLIENT REP. :
DATE :	DATE :

PIPING CONTRACTOR'S
COMPANY LOGO

WELDING PROCEDURE SPECIFICATION REGISTER - PIPING

Client: _____   Title: _____   Contract No.: _____

Contractor: _____   Specification: _____ PCR Job No.: _____

Revision No.: _____   Sheet : _____ of _____   Date : _____

No.	WPS No.	Rev. No	Supporting PQR No.	Welding Process	Material Specification and Grade	Joint Preparation	Qualified Position	Qualified Diameter	Qualified Thickness Range	Minimum Preheating	Maximum Interpass Temperature	PWHT	Impact Test	F-No.	Welding Consumables	Remarks

	Contractor	Third Party	Client
Signature			
Name			
Position			
Date			

PCR-QAF-18

PROCEDURE QUALIFICATION RECORD

A. GENERAL

1. PQR NO. :   TEST DATE :   WELDING CODE / STD :
2. WELDING PROCESS (ES) : SMAW ___ SAW ___ GMAW ___ GTAW ___ FCAW-GS ___
3. EQUIPMENT TYPE : MANUAL ___ SEMI-AUTO ___ MACHINE ___ AUTO ___
4. TYPE OF TEST JOINT :
5. SUPPORTING WPS NO : (1) ___ (2) ___ (3) ___

B. BASE METAL & WELDING CONSUMABLE

1. MATERIAL SPECIFICATION : ___ TO ___
   MATERIAL TYPE / GRADE : ___ TO ___
2. FILLER METAL SPECIFICATION :
   ( ) ___ MFG / BRAND ___
   ( ) ___ MFG / BRAND ___
3. FILLER METAL CLASSIFICATION :
   ( ) ___ A NO. ___ F NO. ___
   ( ) ___ A NO. ___ F NO. ___
4. WELDING FLUX SPECIFICATION : FLUX / ELEC. COMBINATION :
5. WELDING FLUX CLASSIFICATION : MFG / BRAND :
6. TUNGSTEN ELECTRODE DIA. : TYPE : ___ CUP/ORIFICE SIZE :
7. SHIELDING GAS TYPE (S) : ___ MIX % ___ FLOW ___ LPM
8. PURGING GAS TYPE (S) : ___ MIX % ___ FLOW ___ LPM
9. TRAILING GAS TYPE (S) : ___ MIX % ___ FLOW ___ LPM

C. ELECTRICAL

1. WELDING CURRENT & POLARITY :
   ( ) : DC ___ AC ___ POSITIVE ___ NEGATIVE ___ PULSED ___
   ( ) : DC ___ AC ___ POSITIVE ___ NEGATIVE ___ PULSED ___
2. MODE OF METAL TRANSFER FOR GMAW : SHORT CIRCUIT : ___ OTHER : ___
3. WIRE FEED SPEED RANGE, cm/min. : (SAW) ___ (FCAW-GS) ___ (GMAW) ___
4. SUPPLEMENTAL FILLER METAL : POWDERED : ___ GRANULAR : ___ WIRE : ___

D. TECHNIQUE

1. SINGLE OR MULTIPLE ( ) S ___ M ___ STRING ___ WEAVE ___ SPLIT LAYER ___
   PASS USING STRING ( ) S ___ M ___ STRING ___ WEAVE ___ SPLIT LAYER ___
   WEAVE OR SPLIT LAYER : ( ) S ___ M ___ STRING ___ WEAVE ___ SPLIT LAYER ___
2. MAXIMUM WEAVE WIDTH : MANUAL ___ MM SEMI-AUTO ___ MM MACHINE ___ MM
3. VERTICAL WELD PROGRESSION : (FCAW-GS) UPWARD ___ DOWNWARD ___
   ( ) UPWARD ___ DOWNWARD ___
4. POSITION OF TEST WELD : SMAW ___ GMAW ___ FCAW-GS ___ GTAW ___ SAW ___
5. NO. OF ARCS : - ELEC. SPACING: LONGITUDINAL ___ LATERAL ___ ANGLE ___ TRAVEL DIRECTION ___
6. ROOT WELD BACKING : NO ___ YES ___ BACKING MATERIAL ___ WELD METAL ___
7. STICK OUT LENGTH : ___ MM ORIFICE / CUP SIZE : ___ - CONTACT TUBE TO WORK DISTANCE : ___ MM

E. SUPPLEMENT

1. TREATMENT OF BACKSIDE OF ROOT WELD : NONE ___ GRIND ___ GOUGE ___ BRUSH ___ OTHERS ___
2. PREHEAT TEMP. OF TEST PIECE : ___ MAX. INTERPASS TEMP. OF TEST PIECE : ___
3. METHOD OF PREHEAT : ___ METHOD OF TEMP. CHECK : ___
4. PWHT : ___ REPORT NO.: ___
5. CHARPY IMPACT TESTS : ___ REPORT NO.: ___ MIN. (SINGLE) : ___ AVE.: ___
6. HARDNESS TEST : ___ REPORT NO.: ___ MAXIMUM HV : ___ HV
7. TENSILE TEST : ___ REPORT NO.: ___
8. BEND TEST : ___ REPORT NO.: ___
9. MACRO ETCH ANALYSIS : ___ REPORT NO.: ___
10. NDT : ___ REPORT NO.: ___
11. CHEMICAL COMPOSITION : ___ REPORT NO.: ___
12. FERRITE COUNT : ___ REPORT NO.: ___

CONTRACTORS REP :	CLIENT REP :	THIRD PARTY REP :
___	___	___
DATE :	DATE :	DATE :

PCR-QAF-19

PROCEDURE QUALIFICATION RECORD		

PQR NO.: _____   FOR WPS NO.: _____   DATE: _____

JOINT DETAILS	BEAD SEQUENCE

TEST POSITION   : _____
TEST MATERIAL
SPECIFICATION / GRADE   : _____
HEAT NOS   : _____
TEST MATERIAL THICKNESS   : _____
PLATE SIZE   : _____
FILLER METAL MFG/ BRAND   : _____
BATCH NO.   : _____
ROOT WELD BACKING :   : _____
BACKING MATERIAL :   : _____
MAXIMUM WELD WEAVE WIDTH   : _____
(FCAW-GS)   :   (OTHER) : _____
GAS SHIELDING   : TYPE _____ FLOW _____ LPM
GAS PURGING   : TYPE _____ FLOW _____ LPM

MINIMUM PREHEAT TEMPERATURE   : _____
MAXIMUM INTERPASS TEMPERATURE   : _____
PWHT   : _____
NDT   : _____
IMPACT TESTS   : _____
HARDNESS SURVEY   : _____
MACROETCH + PHOTO   : _____
TENSILE TEST   : _____
BEND TEST   : _____
CHEMICAL COMPOSITION   : _____
FERRITE COUNT   : _____
OTHERS **
TUNGSTEN ELEC. TYPE/SIZE   : _____

WELDING PARAMETERS															
LAYER NO.	PASS NO.	PROCESS	FILLER METAL CLASSIFICATION	FILLER METAL DIAMETER (mm)	VERTICAL PROGRESSION	CURRENT TYPE & POLARITY	AMPERAGE (A)	VOLTAGE (V)	TRAVEL SPEED (mm/min.)	HEAT INPUT (kJ/mm)	STRING	WEAVE	SPLIT LAYER	WEAVE WIDTH (mm)	INTERPASS TEMP. (°C)

CONTRACTORS REP :	CLIENT REP :	THIRD PARTY REP :
_____	_____	_____
DATE :- _____	DATE :- _____	DATE :- _____

PCR-QAF-19   Page 2 of 4

PROCEDURE QUALIFICATION RECORD

MECHANICAL AND NON-DESTRUCTIVE RECORD

PQR NO.: _____ FOR WPS NO . : _____ PAGE : _____

BASE METAL : SECT. 1 _____ SECT. 2 _____
HEAT NO. : SECT. 1 _____ SECT. 2 _____
FILLER METAL MFG/BRAND : _____

NON DESTRUCTIVE TESTS

TYPE		GOVERNING CODE			TEST RESULT		REPORT ATTACHED	
RT	UT & MPI	AWS	ANSI	ASME	ACCEPT	REJECT	YES	NO

VISUAL REPORT NO.: _____
RT / UT / MPI REPORT NO.: _____

TENSILE TEST

SPECIMEN NO.	FINISHED WIDTH (mm)	FINISHED THKS (mm)	MEASURED AREA (mm^2)	ULTIMATE LOAD (kN)	ULTIMATE STRESS (N/mm^2)	FAILURE LOCATION

BEND TEST

SIDE BEND	RESULT	SIDE BEND	RESULT

MACRO-ETCH

SPECIMEN NO.	REMOVAL LOCATION	FUSION		POROSITY - SLAG - LOP	
		ACCEPT	REJECT	ACCEPT	REJECT

OTHERS TEST

TYPE	REQUIRED BY			TEST RESULT		REPORT ATTACHED	
	CODE	STD	SPEC.	ACCEPT	REJECT	ACCEPT	REJECT

WE CERTIFY THAT THE STATEMENTS IN THIS RECORD ARE CORRECT AND THAT THE TEST WELDS WERE PREPARED, WELDED AND
TESTED IN ACCORDANCE WITH THE REQUIREMENTS OF WPS NO. : _____
AND _____ CODE FOR WELDING.

ABOVE TEST CONDUCT BY : _____

LABORATORY REPORT TEST NO. : _____

CONTRACTORS REP :	CLIENT REP :	THIRD PARTY REP :
_____	_____	_____
DATE :- _____	DATE :- _____	DATE :- _____

PCR-QAF-19

PIPING CONTRACTOR'S COMPANY LOGO	**WELDING PROCEDURE QUALIFICATION TEST RECORD**

WPS No. :	PQR No. :	Test Coupon No. :	Code / Standard :

Weld Groove / Application :

Process(es)	Base Material : Size and Thickness : Heat No. :	PWHT : Yes [ ] No [ ]	Consumable : Batch No. :
Position / Direction	Treatment of Backside of Root Weld : None/Grind/Gouge/Others	Preheat Temperature :	Gas Shield / Gas Purge :	Gas Flow Rate :

Joint Details :

Weld Profile :

Pass No.	Welding Process	ELECTRODE				Width	Height	Amps	Volts	Travel Speed mm / min	Heat Input kJ / mm	Interpass Temp. °C
		Size (mm)	Classification	Brand	Polarity							

Prepared by : Date :	Witnessed / Reviewed by : Date :

PIPING CONTRACTOR'S COMPANY LOGO

PROCEDURE QUALIFICATION RECORD REGISTER

Client:	Title:	Contract No.:
Contractor:	Specification:	PCR Job No.:
Revision No.:	Sheet: _____ of _____	Date:
	Minimum Design Temperature:	

PQR No.	Rev. No.	Date of Test	Welding Process	Material Specification & Grade	W.T. mm	Dia. mm	Material Group Qualified	Thickness Min - Max	Diameter Mix - Max	PWHT	Application

	Contractor	Third Party	Client
Signature			
Name			
Position			
Date			

PIPING CONTRACOR'S COMPANY LOGO	WELDER & WELDING OPERATOR QUALIFICATION TEST CERTIFICATE (AWS/ASME)

NAME : _____   ID NO. : _____
I/C NO. : _____   WPS NO.: _____
STATUS : _____

	DETAILS AS TESTED	RANGE QUALIFIED
PROCESS	: _____	_____
PROCESS TYPE	: _____	_____
POSITION	: _____	_____
WELD PROGRESSION	: _____	_____
BASE METAL	: _____	_____
P No. / Type / Group No.	: _____	_____
Nominal Diameter	: _____	_____
Thickness	: _____	_____
BACKING MATERIAL	: _____	_____
BACKING GAS	: _____	_____
FILLER METAL		
Trade Name / Diameter	: _____	_____
Classification	: _____	_____
Specification / SFA No.	: _____	_____
F No.	: _____	_____
EQUIVALENT TRANSFER MODE	: _____	_____
GAS / FLUX TYPE	: _____	_____
ELECTRICAL CHARACTERISTICS		
Current / Polarity	: _____	_____

REMARKS :

TEST RESULT
STATE **'ACCEPT'** OR **'NOT REQUIRED' (N/R)**

NON-DESTRUCTIVE TESTS		
TEST	RESULT	REPORT NO.
VISUAL		
RADIOGRAPHIC		
ULTRASONIC		
MAGNETIC PARTICLE		
PENETRANT		

DESTRUCTIVE TESTS		
TEST	RESULT	REPORT NO.
MACRO/MICRO		
FILLET WELD BREAK		
SIDE BEND		
ROOT BEND		
FACE BEND		

THE STATEMENTS IN THIS CERTIFICATE ARE CORRECT. THE TEST WELD WAS PREPARED, WELDED AND TESTED IN ACCORDANCE WITH THE REQUIREMENTS OF

	CONTRACTOR	THIRD PARTY	CLIENT
Signature			
Name			
Position			
Date			

WELDER'S ID CARD - FRONT PAGE

PIPING CONTRACTOR'S COMPANY LOGO

WELDER NAME   :

NRIC   :

WELDER ID NO.   :

WELDER
PHOTO

WELDING PROCESS:

PCR-QAF-22

WELDER'S ID CARD - BACK PAGE

PROCESS	P	WPS NO	DATE OF TEST	RANGE OF QUALIFICATION	MATERIAL	FILLER RANGE

PIPING CONTRACTOR'S COMPANY ADDRESS:

TEL :   FAX :

WEB SITE : www.

PIPING CONTRACTOR'S COMPANY LOGO		WELDER QUALIFICATION RENEWAL ENDORSEMENT RECORD	

Welder's Name		ID No.		Process	
Qualification Renewed up to	Signature/Stamp of PCR's Welding Engr.	Signature/Stamp of TPI (If applicable)		Reference Report/s	

PIPING CONTRACTOR'S COMPANY LOGO

WELDING OVEN LOG BOOK

Client: _____ Title: _____ Contract No.: _____

Contractor: _____ Specification: _____ PCR Job No.: _____

S/N	ELECTRODE SPECIFICATION			1st BAKING PARAMETERS				1st HOLDING PARAMETERS				ELECTRODE RETURN DETAILS			2nd BAKING PARAMETERS				2nd HOLDING PARAMETERS				SIGNATURE WITH REMARKS
	CLASS NO.	SIZE	QUANTITY	DATE	OVEN TEMP (°C)	TIME IN	TIME OUT	DATE	OVEN TEMP (°C)	TIME IN	TIME OUT	TIME	QUANTITY (KG)	DATE	OVEN TEMP (°C)	TIME IN	TIME OUT	DATE	OVEN TEMP (°C)	TIME IN	TIME OUT	

	Contractor		Contractor
Signature			
Name			
Position	Supervisor/Foreman		QC Engineer/QC Inspector
Date			

PIPING CONTRACTOR'S
COMPANY LOGO

WELDING CONSUMABLES ISSUE BOOK

Client:

Contractor:

Title:

Specification:

Contract No.:

PCR Job No.:

Date	Project No.	Specification	Size	Quantity Issued (Kg)	Authorized Supervisor	Quantity Returned (Kg)	Supervisor Signature	QC Inspector	Remarks	WPS No.	Welder ID

PIPING CONTRACTOR'S COMPANY LOGO		NDE REQUEST	

Client :	Title :		Contract No.:
Contractor :			PCR Job No.:
Request No.:	Specification :		Date :

Sl. No.	Drawing No. / Line No.	Sht. No.	Weld No.	Welder No.	Size	Sch/Thk	NDE Method	Remarks

Comments/Sketch

Requested by : Contractor	Received by : Contractor
Signature :	Signature : ....
Name :	Name : ....
Position : QC Inspector	Position : NDT Coordinator
Date :	Date : ....

Piping Contractor's Logo	Piping Contractor's Company Name / NDT Company's Name

Client Name

			Position	Interpretation	Evaluation
Project	:				
Line No.	:				
Identification	:				
Inspection Code	:				
Material	:				
Welder No.	:				
Process	:				
Extent of Examination	:				
Interpreted By	:				
Date	:				

PCR-QAF-27

PIPING CONTRACTOR'S COMPANY LOGO

NDT TRACKING SUMMARY

Details from NDE Request															Initial RT Report		R1 RT Details from RT Report						R2 RT Details from RT Report						SC1 RT Details from RT Report						
Sl. No.	Area	Dwg. No.	Sht. No.	Size	Fluid	Line No.	Specification	Joint No.	Material Spec.	Diameter	Thickness	Welding Process	Welder No.	RT Requested Date	RT Date	RT Report No.	Repair R1 Attended Date	NDE Request for Repair	RT Taken Date	RT Report No.	RT Result by NDT Subcon	RT Result by Client	Repair R2 Attended Date	NDE Request for Repair R2	RT Taken Date	RT Report No.	RT Result by NDT Subcon	RT Result by Client	Cutout (SC1) Attended Date	NDE Request for Cutout (SC1)	RT Taken Date	RT Report No.	RT Result by NDT Subcon	RT Result by Client	REMARKS

PCR-QAF-28

PIPING CONTRACTOR'S
COMPANY LOGO

PUNCH LIST

Client:		Title:		Contract No.:	
Contractor:				PCR Job No.:	
System Reference:		Package No.:		Date:	

Item No.	Drawing No.	Observations/Findings	Category*	Punch Killing	
				Signature	Date

	Contractor	Contractor	Client
Signature			
Name			
Position	Engineer/Supervisor	QC Engineer/Inspector	
Date			

* A - To be completed prior to flushing/testing   B - To be completed after flushing/testing   C - To be completed before commissioning

PCR-QAF-29

Page 1 of 1

PIPING CONTRACTOR'S COMPANY LOGO

PUNCH LIST TRACKING REGISTER

Date:

Sl. No.	Discipline	Total Punchlist					Completed by Construction (PCR)					Closed out by Client					Open					Remarks
		"A"	"B"	"C"	"D"	Total	"A"	"B"	"C"	"D"	Total	"A"	"B"	"C"	"D"	Total	"A"	"B"	"C"	"D"	Total	
1	e.g.: Civil																					
2	Structure																					
	Total	0																				
	%																					

PCR-QAF-30

TEST PACKS TRACKING REGISTER

PIPING CONTRACTOR'S COMPANY LOGO

Date:

Sl. No.	System / Service	System No.	TOTAL					PNEUMATIC TESTING...					

The main register table contains the following column groups and sub-columns:

TOTAL
- Total Quantity
- Pretest Walk down Completed
- NDE Cleared / Released for Testing
- Testing Completed
- Reinstatement Completed

HYDRO TESTING
- Total Quantity
- Pretest Walk down Completed
- NDE Cleared / Released for Testing
- Testing Completed
- Reinstatement Completed

PNEUMATIC TESTING
- Total Quantity
- Pretest Walk down Completed
- NDE Cleared / Released for Testing
- Testing Completed
- Reinstatement Completed

SERVICE TESTING
- Total Quantity
- Pretest Walk down Completed
- NDE Cleared / Released for Testing
- Testing Completed
- Reinstatement Completed

VISUAL INSPECTION ONLY
- Total Quantity
- Pretest Walk down Completed
- NDE Cleared
- Testing Completed
- Reinstatement Completed

Remarks

Summary table:

	Total Quantity	Pretest Walk down Completed	NDE Cleared / Released for Testing	Testing Completed	Reinstatement Completed
Hydro Testing					
% Completed					
Pneumatic Testing					
% Completed					
Service Testing					
% Completed					
Visual Testing					
% Completed					
Total					
% Completed					

PCR-QAF-31

PROJECT FINAL DOCUMENTATION APPROVAL--HANDOVER STATUS

Volume	Description	Responsibility	MEB Status			Client Status				Scanning Status				Bookmark Status				Handover to Client						REMARKS
			Total Books	Total Books Compiled	BALANCE TO COMPILE	Under Review	Reviewed	Approved	BALANCE TO BE APPROVED	Total Books to Scan	Completed	BALANCE	PIC	Total Books to Bookmark	Completed	BALANCE	PIC	Master Books	Total CDs	Completed	BALANCE	Handover Date	Transmittal Ref. No.	
VOL. 1	Design Data Dossier											# 0												
VOL. 2	Equipment Technical Dossier																							
VOL. 3	Equipment QA/QC Dossier																							
VOL. 4	Construction QA/QC Dossier																							
VOL. 4A	Fabrication Structural Dossier																							
VOL. 4B	Installation QA/QC Dossier																							
VOL. 5	Commissioning Dossier																							
VOL. 6	Start-up Manual																							
VOL. 7	As-Built Drawing																							
VOL. 8	Others																							
VOL. 8A	Regulatory & Compliances																							
VOL. 8B	Material Dossier																							
VOL. 8C	Project Management																							
VOL. 8D	HSE Dossier																							
VOL. 8E	Project Closeout Report																							
	Total		0	0	0	0	0	0	0	0	0	0	0	0	0	0		0	0	0	0			

PCR-QAF-32

LIST OF REFERENCE FORMS:

The following formats shall be used for oil and gas piping related works in the projects. Contents of the formats are based on our experience and can be changed to suit the project specific requirements as well as client requirements. In many cases, clients will have their own formats and they will ask us to use their formats. This can be followed. However, care should be taken to ensure that, redundancy in data entering and increased paper works should not arise for the people involved in the project.

Sl. No.	Section No.	Annexure No.	Form No.	Form Title
01	One	Annexure - 1	PCR-QAF-01	3rd Party RFI
02	One	Annexure – 2	PCR-QAF-02	Source Inspection report
03	One	Annexure – 3	PCR-QAF-03	Material receipt inspection and test report
04	One	Annexure – 4	PCR-QAF-04	List of material required identification
05	One	Annexure – 5	PCR-QAF-05	List of authorized personnel for withdrawing material
06	One	Annexure – 6	PCR-QAF-06	Stock card
07	One	Annexure – 7	PCR-QAF-07	Store issue note
08	One	Annexure – 8	PCR-QAF-08	Non-conformance report
09	Two	Annexure – 9	PCR-QAF-09	Document Transmittal
10	Two	Annexure – 10	PCR-QAF-10	Fit-up and visual inspection record (Piping)
11	Two	Annexure – 11	PCR-QAF-11	Dimensional report
12	Two	Annexure – 12	PCR-QAF-12	Weld summary sheet (Piping)
13	Two	Annexure – 13	PCR-QAF-13	Release note for painting
14	Two	Annexure – 14	PCR-QAF-14	Blasting and Painting inspection report
15	Two	Annexure – 15	PCR-QAF-15	Painting or coating final inspection report
16	Two	Annexure – 16	PCR-QAF-16	Internal packing list
17	Three	Annexure – 17	PCR-QAF-17	Welding procedure specification
18	Three	Annexure – 18	PCR-QAF-18	Welding procedure specification register
19	Three	Annexure – 19	PCR-QAF-19	Welding procedure qualification record
20	Three	Annexure – 20	PCR-QAF-20	Welding procedure qualification record register
21	Three	Annexure – 21	PCR-QAF-21	Welder and welding operator qualification test certificate (AWS/ASME)
22	Three	Annexure – 22	PCR-QAF-22	Welders ID card
23	Three	Annexure – 23	PCR-QAF-23	Welder qualification
24	Three	Annexure – 24	PCR-QAF-24	Welding oven log book
25	Three	Annexure – 25	PCR-QAF-25	Welding consumable issue book
26	Four	Annexure – 26	PCR-QAF-26	NDT Request form
27	Four	Annexure – 27	PCR-QAF-27	NDT Film identification / wrapping form
28	Four	Annexure – 28	PCR-QAF-28	NDT Tracking Summary
29	Six	Annexure – 29	PCR-QAF-29	Punch list
30	Six	Annexure – 30	PCR-QAF-30	Punch List Tracking Summary
31	Six	Annexure – 31	PCR-QAF-31	Test Packs Reporting Summary
32	Seven	Annexure - 32	PCR-QAF-32	Project Final Documentation Handover – Approval Tracking.

Printed and bound by CPI Group (UK) Ltd, Croydon, CR0 4YY

22/10/2024

01777637-0006